Advances in Growth Curve and Structural Equation Modeling

Ratan Dasgupta
Editor

Advances in Growth Curve and Structural Equation Modeling

Topics from the Indian Statistical Institute—Proceedings 2017

Springer

Editor
Ratan Dasgupta
Theoretical Statistics
and Mathematics Unit
Indian Statistical Institute
Kolkata, West Bengal, India

ISBN 978-981-13-0979-3 ISBN 978-981-13-0980-9 (eBook)
https://doi.org/10.1007/978-981-13-0980-9

Library of Congress Control Number: 2018943729

This Springer imprint is published by the registered company Springer Nature Singapore Pte Ltd.
The registered company address is: 152 Beach Road, #21-01/04 Gateway East, Singapore 189721, Singapore

Preface

A growth curve is an empirical model of the evolution of certain characteristic of interest over time. Growth curve models are widely studied in different branches of science. The present volume on growth curve model (GCM) is a culmination of the talks given at the workshop on the topic held during 23–24 February 2017 at the Indian Statistical Institute, Giridih, where growth experiments are going on in the farmland.

Professor Jayanta K. Ghosh, an eminent academician and a former director of the Indian Statistical Institute, passed away on September 30, 2017. He was one of the inspiring forces in conducting growth experiments in Giridih over years. The present volume is dedicated to his memory.

This workshop proceeding 'Advances in Growth Curve and Structural Equation Modeling: Proceedings 2017—Topics from the Indian Statistical Institute' presents some ideas about the research works, both theoretical and applied, on growth curve models, which are going on by the scientists of Indian Statistical Institute in different branches of science over years. Growth curve models are used with great effectiveness in many areas of developmental research. I am thankful to the readers as the previous three volumes on GCM—Advances in Growth Curve Models: Topics from the Indian Statistical Institute, 2013, Growth Curve and Structural Equation Modeling: Topics from the Indian Statistical Institute, 2015, and Growth Curve Models and Applications, Indian Statistical Institute, Giridih, India, March 28–29, 2016 (2017)—are well accepted by the scientific community.

We invited for contribution to this workshop proceeding and further invited the participants of the workshop to submit more than one paper, if possible for the proceedings. All the papers were peer-reviewed. The result is a compilation of 11 research papers presented in this volume.

The endeavor will be considered successful, if this can give some idea about solving theoretical and practical problems in this broad area of GCM in which many researchers in different branches of science are interested in.

Kolkata, India
October 2017

Ratan Dasgupta

Contents

Editor and Contributors

About the Editor

Prof. Ratan Dasgupta is Senior Professor of Statistics at the Indian Statistical Institute, Kolkata. His research interests include nonparametric statistics, rates of convergence in central limit theorem, the application of statistics to industrial quality control, sociology, education, agriculture, biology, physics, and other natural sciences. He has been at the forefront of promoting the theory and the applications of growth curve modeling, and his knowledge, expertise and extensive publication record on the topic, together with his outstanding theoretical skills make him uniquely qualified to edit this volume.

Contributors

Premananda Bharati Biological Anthropology Unit, Indian Statistical Institute, Kolkata, India

Susmita Bharati Sociological Research Unit, Indian Statistical Institute, Kolkata, India

Koushik Biswas Regional Institute of Ophthalmology, Medical College Hospital, Kolkata, India

Samarjit Das Economic Research Unit, Indian Statistical Institute, Kolkata, India

Debabrata Dash Department of Biochemistry, Institute of Medical Sciences, Banaras Hindu University, Varanasi, India

Ratan Dasgupta Theoretical Statistics and Mathematics Unit, Indian Statistical Institute, Kolkata, West Bengal, India

Monoranjan Ghose Agricultural and Ecological Research Unit, Indian Statistical Institute, Kolkata, India

Hema Gupta (Joshi) Agricultural and Ecological Research Unit, Indian Statistical Institute, Kolkata, India; Department of Botany, Visva-Bharati, Bolpur, Santiniketan, West Bengal, India

Nripes K. Mandal Calcutta University, Kolkata, India

Manisha Pal Calcutta University, Kolkata, India

Manoranjan Pal Economic Research Unit, Indian Statistical Institute, Kolkata, India

Sattwik Santra Centre for Training and Research in Public Finance and Policy, Centre for Studies in Social Sciences, Kolkata, India

Bikas K. Sinha Indian Statistical Institute, Kolkata, India

T. S. Vasulu Indian Statistical Institute, Kolkata, India

Longitudinal Studies on Mathematical Aptitude and Intelligence Quotient of North Eastern Tribes in Tripura

Ratan Dasgupta

Abstract Longitudinal studies are conducted on mathematical aptitude and intelligence quotient on North Eastern tribes in Tripura over successive interviews in a time span of more than 3 years, viz., 20 September 2011–28 November 2014. Analyzed longitudinal data indicate that both mathematical aptitude and intelligent quotient scores exhibit fluctuations over time and have upward trend immediately after first interaction with the interviewer, before stabilizing at a level slightly below the peak value of scores. Average level of mathematical aptitude is low, although the level of intelligence quotient score is comparatively high. Growth curves under different setups are estimated to infer about the status of tribal education and lifestyle. The score status is seen to be improving over time, although associated with mild fluctuations. Proliferation rates of different scores are estimated under different assumptions. In general, the proliferation rates reach stability towards the end of curves for large values of time. Postulating a simple model of association in scores over time based on martingales, we examine the fluctuation of scores. Excessive deviation results for martingales are derived. Under certain conditions on the martingale $\{M_i : 1 \leq i \leq n\}$, the excessive deviation $P(\max_{1\leq i\leq n} |M_i| \geq \lambda n^{1/2})$ is seen to be $O(e^{-\frac{\lambda^2}{2}(1+o(1))})$; $\lambda \to \infty$. This is similar to the tail probability of normal distribution. Deviation of observations from response curve may be compared with normal deviate to detect the presence of assignable causes.

Keywords Lifestyle status · Mathematical aptitude · IQ · Proliferation rate · Martingale · Excessive deviation · Kokborok

MS Subject Classification: 62P25 · 60G20

R. Dasgupta (✉)
Theoretical Statistics and Mathematics Unit, Indian Statistical Institute,
203 B.T. Road, Kolkata 700108, India
e-mail: rdgupta@isical.ac.in; ratandasgupta@gmail.com

R. Dasgupta (ed.), *Advances in Growth Curve and Structural Equation Modeling*,
https://doi.org/10.1007/978-981-13-0980-9_1

1 Introduction and Sketch of the Work

Studies on mathematical aptitude and intelligence quotient in different ethnic groups are of interest. These scores can successfully be utilized to predict the potential of individuals in different fields of activities at a later stage of life. For a group of people, high level of the scores may be interpreted as an indication of education and general development. Studies conducted in a vulnerable section of society indicated that exploitation and education are inversely related, e.g., see p-321 of: http://dspace.vidyanidhi.org.in:8080/dspace/bitstream/2009/4402/10/OSM-1993-041-9.pdf.

The whole of the North East has been categorized as industrially backward. See, e.g., http://planningcommission.nic.in/reports/publications/pub_neregion.pdf.

In terms of development, many of the North Eastern states lags behind the rest of the country and are not able to come to the forefront of education, see e.g., http://timesofindia.indiatimes.com/india/7-of-8-Northeast-states-lag-behind-average-India-income/articleshow/5187537.cms.

We study the above two traits, viz., Mathematical aptitude and IQ in tribal individuals in northeastern part of Tripura state with 15 simple basic questions related to mathematical ability and 7 questions on general knowledge and reasoning. In the beginning of the investigation, the questions in interviews were asked and explained verbally by the interviewer. Some answers were provided by tribal individuals in writing and others were recorded by interviewer. In the later stage of investigation, written questionnaires were provided, the answers were given by the tribal individuals in between the blank spaces of printed questions. Total score and the component scores on IQ and Math aptitude are standardized in the range [0, 100] before plotting.

Analyzed data indicate an upward trend immediately after the first interaction of individuals with the interviewer, before stabilizing at a level slightly below the peak. This may be due to the fact that increase in score can be achieved, in general, by coaching people in the way of answering/explaining items on intelligent test. Coaching can create exceptional increase of a temporary nature. Also, any increase in scores tends to level out after several coaching sessions and ultimately that fades with time, see Duminy et al. (1991).

A similar phenomenon may happen when the tribal individuals are discussing lifestyle with interviewer in the course of recording answers to such questions on lifestyle and IQ posed to them for the first time. Here, coaching can be compared with interaction of tribal individuals with the interviewer when subject matter is explained in a lucid manner. Impact of this fruitful interaction is reflected in the scores of the next interview.

Language of instruction for many schools in Tripura is Bengali, while the mother tongue of a large segment of students is *Kok Borok* in tribal belt. Linguistic handicaps are experienced where medium of school instructions is not the mother tongue. Unless children can acquire complete facility in the language at school, inevitably they are retarded in reasoning as well as attainment. Often they can learn to speak the language fairly accurately, but they do not think naturally in it; instead they tend to revert to the less efficient but more deeply rooted mother tongue.

The individuals are interviewed at different places of congress like students in different schools, marketplaces, church, and villages after stratification. Selection of individuals for interview is made by simple random sampling without replacement at final stage. Non-response of a unit is remedied by selecting an adjacent available unit.

In this pilot survey, data are collected in the way of personal interviews, administering tests and filling up the questionnaire by selected individuals. Sometimes resistance to fill up the questionnaire was faced from skeptic elements, motif and ultimate goal of such studies were questioned and monetary angle out of the results in conducted survey was prodded into, while working in some interior parts of the state. It appears that affinity of inhabitants of interior hilly region and plane region where modern facilities are available is slowly improving over time with progress of education and lifestyle.

In the next section, we present the results obtained from the data analyzed. The error components in the scores may not be symmetric in all cases.

The individuals may not be highly educated in general, but their receptivity of to new concepts is commendable. Figures based on scores provide insight into the lifestyle and educational status of individuals interviewed.

Fluctuations in scores are modeled by martingale differences. Excessive deviation results for martingale are obtained.

In Sect. 2, we present the results in brief. These are further elaborated in figures with legends and associated explanation given in detail therein. Section 3 discusses the results obtained. Fluctuations of observations around the response curves are explained by martingale theory. While estimating the response curve by nonparametric regression, assume that the differences of the existing data point from regression curve behave like martingale differences as new data are gathered from recent interview to modify the regression, and those successive martingale differences constitute a martingale. Maximal fluctuation of the martingales is of interest.

Excessive deviation results for maximal fluctuations are proved in the Appendix. Under assumptions on moment bounds, those are not so stringent on martingale differences, and deviation probabilities of maximal fluctuations in martingales are seen to decay with order higher than any polynomial power of the boundary value. Since the total score and the component scores on IQ and Math aptitude are standardized in the range [0, 100], we may assume that the martingale differences are bounded. In such a situation, for the martingales $\{M_i : 1 \leq i \leq n\}$, we compute the excessive deviation probability as $P(\max_{1\leq i\leq n}|M_i| \geq \lambda n^{1/2}) = O(e^{-\frac{\lambda^2}{2}(1+o(1))})$; $\lambda \to \infty$, which is similar to the tail probability of normal distribution. High oscillations of observations from the response curve may be assessed in terms of normal deviate to identify assignable causes.

In general, the level of IQ is seen to be higher than that of scores in mathematical aptitude. A large segment of the individuals under the scope of study understood the

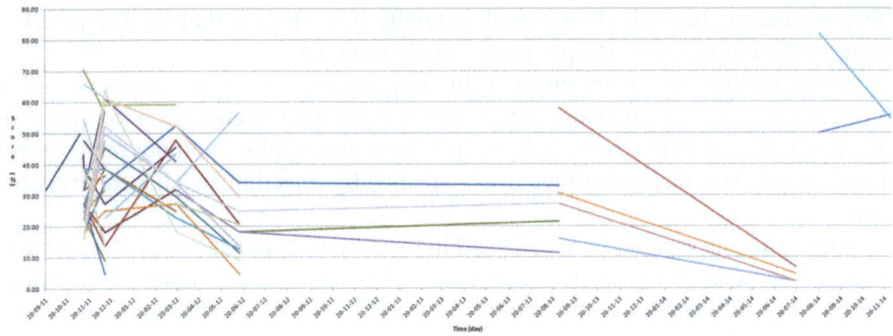

Fig. 1 Longitudinal score (Math plus IQ) of 47 tribal individuals in Tripura. Longitudinal growth curves of total scores for 47 tribal individuals in Mathematics and IQ after scaling are shown in Fig. 1, the maximum possible score is scaled to 100. The number of Math questions is 15, and the number of IQ questions is 7 with a total of 22 questions administered in interviews. Indications of upward trend in second interview scores are possibly due to fruitful interaction with the individuals at the first interview date, on topics at introductory level, before these tests were administered. Elaboration of the subject matter affects different individuals over a considerable period of time, and this phenomenon is apparent in observed peaks. Too many lines make the picture congested. Some individuals appeared only twice, which results in two interview scores over time

aim of the tests and were cooperative with the interviewer on successive occasions of test administration.

2 Results

The survey conducted over a time span of about 3 years revealed that IQ test score in percentage is at a higher level than the Math score percentage. Data analyzed under the assumption that error component is symmetrically distributed and is consistent with other nonparametric methods to estimate the growth curve for total score of Math plus IQ, and Math score alone. Symmetry of error component may not hold for IQ score, expressed in percentage. An upward trend is seen on scores after interaction with the interviewer, indicating receptivity of individuals towards improvement of knowledge. Error component in IQ score may not be symmetric. From the growth curve of Math aptitude and IQ scores, it appears that the education status of tribal individuals is improving over time. Some of the scores seem to attain stability over time towards the far end. Different features of collected data and growth curve related results are further explained in the Figures.

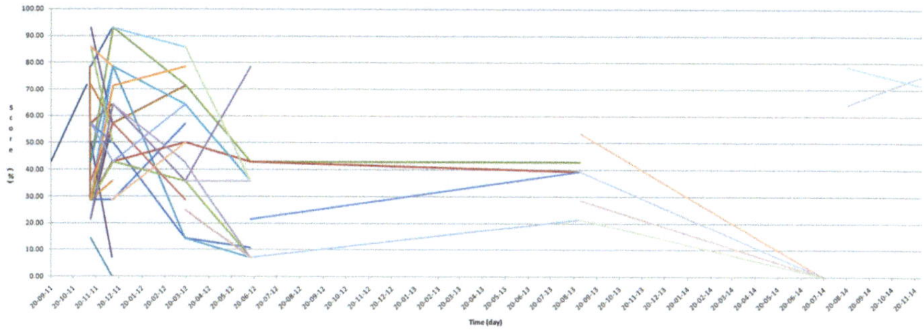

Fig. 2 Longitudinal score (IQ) of 47 tribal individuals. Questions on Math and IQ are separated and scores are also separated into two components and expressed in percentage. The first component IQ score in percentage for longitudinal study is shown in Fig. 2. Patterns in these growth curves are similar to that of Fig. 1, the level of IQ scores has relatively higher elevation than total score expressed in percentage

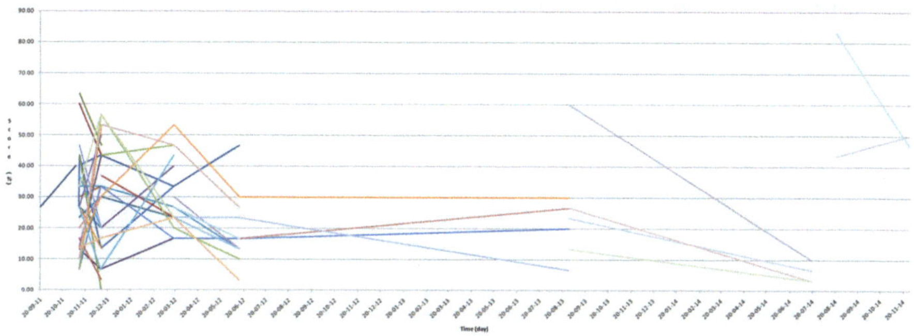

Fig. 3 Longitudinal score (Math) of 47 tribal individuals. The second component, Math score in longitudinal study is shown in Fig. 3. Elevation of scores is a bit lower that earlier two graphs, and pattern of the graph is the same with former two graphs. Overall performance in Mathematics seems to be poor compared to that in IQ

3 Discussions

Investigation on the longitudinal aspect of scores in Mathematical aptitude and IQ on a segment of tribal population revealed that although the general level of education is low, improvement over time is observed especially after first exposure to such aptitude tests and interaction with the interviewer. In general, the level of IQ is higher than that of scores in mathematical aptitude. Receptivity of the individuals to new concepts and ideas is commendable. The response curve for the two traits under investigation may be examined under the mild assumption that the errors are martingale. Excessive deviation results for maximal fluctuations of martingales are derived under different moment assumptions. The results are similar to the normal tail probabilities. Probability of high fluctuation of observations from the response

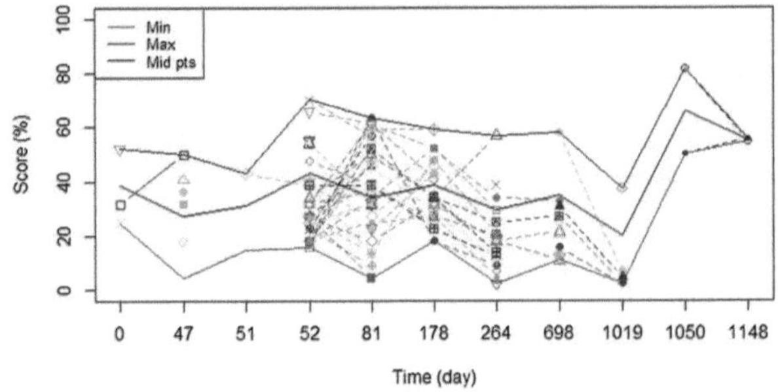

Fig. 4 Band of growth curves (Math plus IQ) of 47 individuals. The band that contains all 47 longitudinal curves on total score is shown by joining the extreme points with straight lines. For each value of time x, the line of y coordinate has shortest length within the band containing all individual curves. The curve joining mid points of the upper and lower band is also shown in Fig. 4. The curve in the middle represents an estimate of overall growth in total scores under the assumption that error components, i.e., deviation of scores from response curve for each time point t have symmetric distribution. Mid-range is sometimes more efficient than the mean for estimating central tendency in a number of distributions with limited range, see, e.g., Rider (1957)

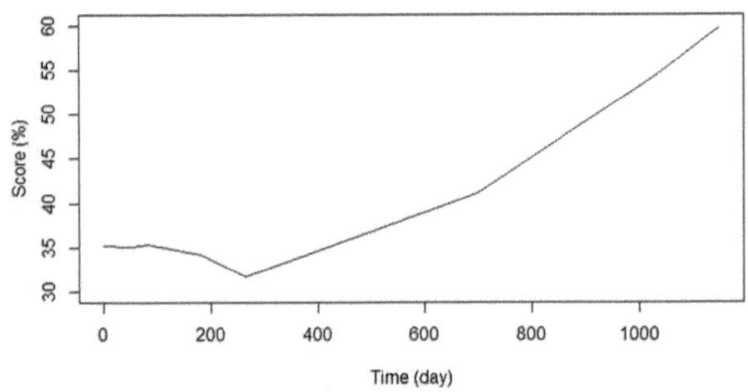

Fig. 5 Lowess on mid points of growth band (Math + IQ) of 47 individuals. A nonparametric growth curve of total score by lowess regression on mid points of the band with $f = 2/3$ is shown in Fig. 5. The curve has initial upward trend for a while, it goes down and reaches a minimum when initial boost from interaction with interviewer possibly erodes. Then the curve shows an upward trend gradually over time. We shall see later that the pattern of the growth curve remains the same, if the median of the scores on days on recording is considered instead of midpoints of the band considered in lowess regression. This is in concordance with the assumption that errors have symmetric distribution for total score

curve may then be compared with normal tail probability to investigate for assignable causes.

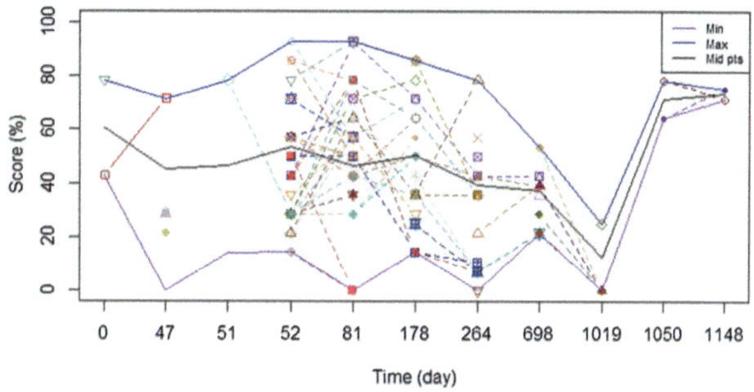

Fig. 6 Band of growth curves (IQ) of 47 individuals. A similar band containing of 47 longitudinal curves on IQ score is shown in Fig. 6. Here again, the curve in the middle represents an estimate of overall growth of IQ score under the assumption that errors have symmetric distribution

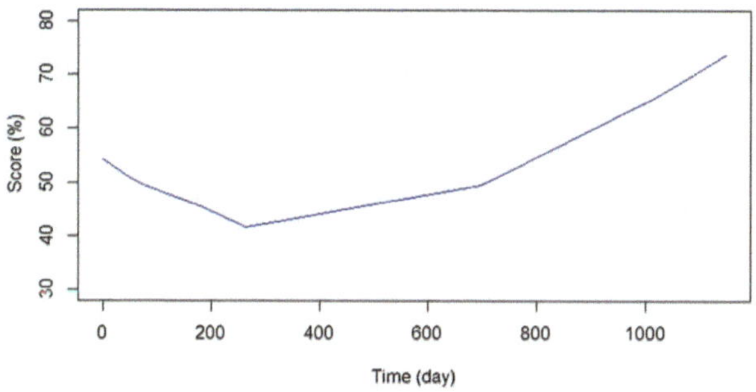

Fig. 7 Lowess on mid points of growth band (IQ) of 47 individuals. A nonparametric growth curve of IQ score by lowess regression on midpoints of the band with $f = 2/3$ is shown in Fig. 7. The curve has initial downward trend and reaches a minimum, then the curve gradually exhibits an upward trend over time. We shall see later in Fig. 26 that the pattern of the graph changes when the assumption of symmetric error is not made, indicating that the assumption of symmetric error component for IQ scores may not hold

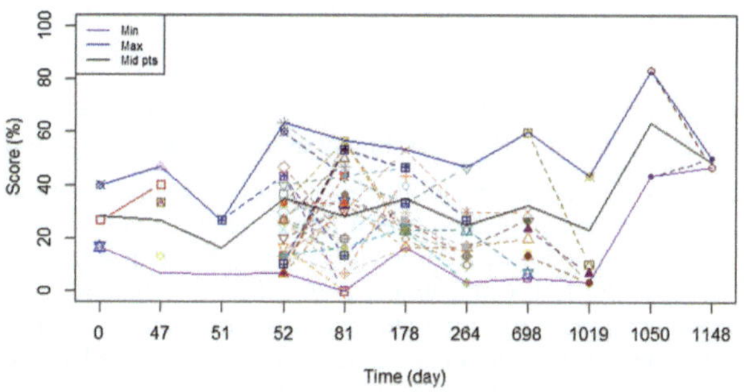

Fig. 8 Band of growth curves (Math) of 47 individuals. The band containing 47 longitudinal curves on mathematics score is shown in Fig. 8. The middle curve joining the midpoints of the band provides an estimate of growth curve in mathematics score

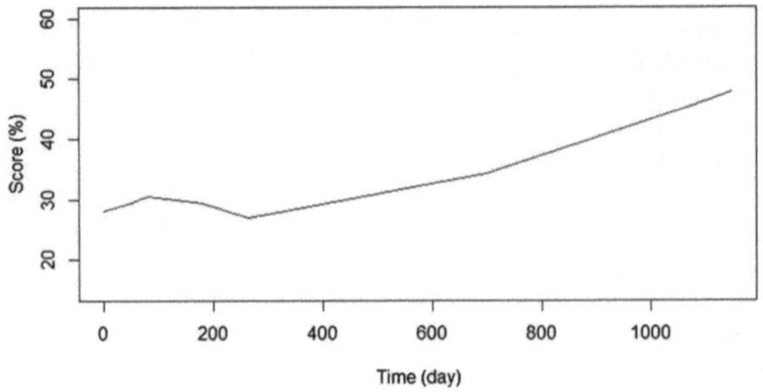

Fig. 9 Lowess on midpoints of growth band (Math) of 47 individuals. A nonparametric growth curve of mathematics score by lowess regression on midpoints of the band with $f = 2/3$ is shown in Fig. 9. The curve has an initial upward trend for a while, then it goes down and reaches a minimum when initial boost from interaction with interviewers possibly erodes; in later part, the curve gradually shows an upward trend over time

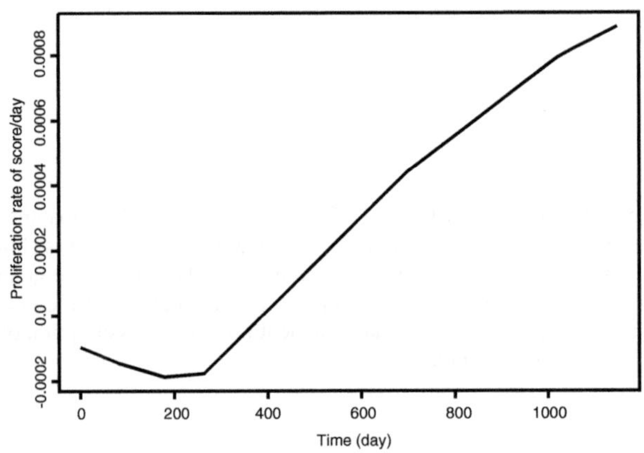

Fig. 10 Proliferation rate of total score: wt. $\exp(-x)$; spline. We compute the proliferation rate based on the growth curve shown in Fig. 5. Proliferation rate $\frac{d}{dt}\log y = \frac{1}{y}\frac{dy}{dt}$ is a scaled version of velocity $\frac{dy}{dt}$. This measure is independent of the choice of unit used in measuring y. For total score y with growth curve computed in Fig. 5 by lowess regression, the proliferation rate is obtained in Fig. 10. The curve has an initial downward trend up to day 190, and then the curve rises gradually, with fast growth towards the end. Computation is based on a technique proposed in Dasgupta (2015), with exponentially decaying weights attached to empirical slopes computed from data pairs at different time points with respect to a fixed time point t of interest. More weights are given to data points near the time t of derivative computation, and less weights to distant time points from t. Weighted mean of these empirical slopes at derivative stage and smooth.spline with spar $= 0.01$ at smoothing stage in SPlus provide proliferation rate at time point t, when divided by y. Instead of considering median or trimmed mean that produces too smooth rates especially towards end of the range, we take weighted average of empirical slopes; with less weight assigned to slopes computed from distant points

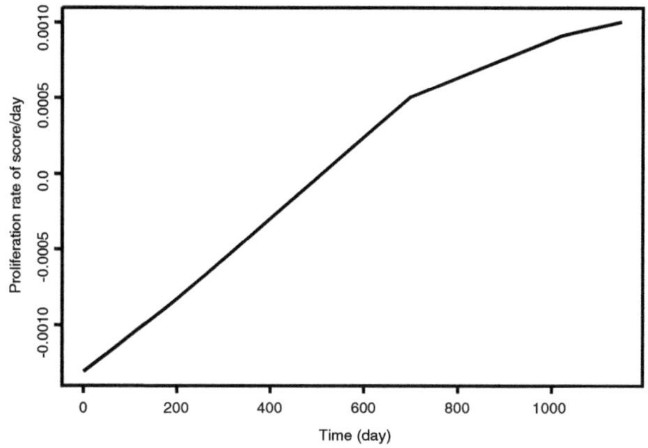

Fig. 11 Proliferation rate of IQ score: wt. exp$(-x)$; spline. Proliferation rate $\frac{d}{dt}\log y$ over 11 time points for IQ score y is computed as in Fig. 10, with exponentially decaying weights attached to empirical slopes computed from data pairs at different time points with respect to a fixed time point t of interest. Weighted mean of the empirical slopes is considered at derivative stage of computation and the package smooth.spline with spar $= 0.01$ at smoothing stage in SPlus provides the proliferation rate at the time point t. A consistent sharp rise in the beginning and then slightly slow rise of rate towards the end is seen in Fig. 11. However, we shall see later that the pattern changes when we consider median instead of mid-values of the band to estimate central tendency. This is because in Fig. 7, the basic input shown for computing proliferation rate here changes to a different pattern when the assumption of symmetric error for IQ is dropped

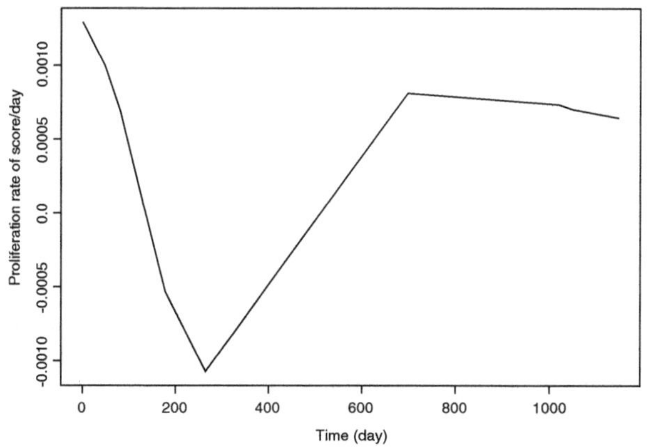

Fig. 12 Proliferation rate of math score: wt. exp$(-x)$; spline. Proliferation rate $\frac{d}{dt}\log y$ over 11 time points for Math score y is computed as in Fig. 10, with exponentially decaying weights attached to empirical slopes computed from data pairs at different time points with respect to a fixed time point t of interest. Weighted mean of the empirical slopes is considered at derivative stage of computation and the package smooth.spline with spar $= 0.0001$ at smoothing stage in SPlus provide the proliferation rate at the time point t. A downward trend in the beginning and then a typical sharp rise towards middle is observed in Fig. 12; the rate slows down towards the end and approaches stability over time

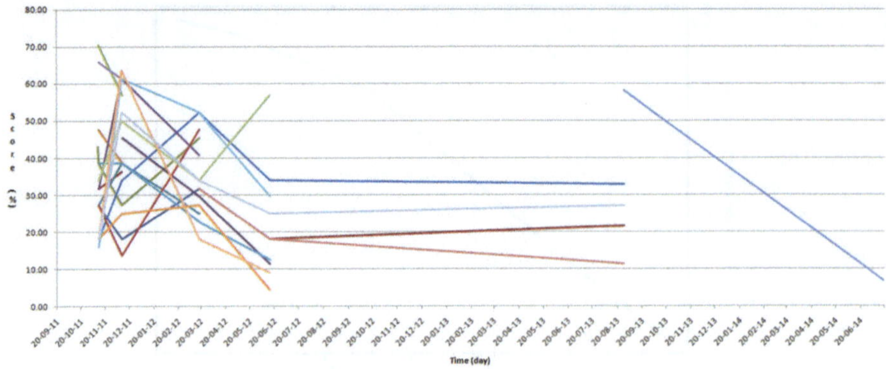

Fig. 13 Longitudinal score (Math plus IQ) of 19 individuals with more than two recordings. Analysis based on longitudinal data with many recordings per individuals has accurate assertions. We now consider only those cases where more observations are available per individuals, as there are some longitudinal growth curves with only two recordings. When tracked over time, 19 individuals had more than two recordings. Total scores of Math plus IQ for these individuals are shown in Fig. 13. Figure 13 is less messy compared to Fig. 1. Upward trend of scores in general are seen after the first interview

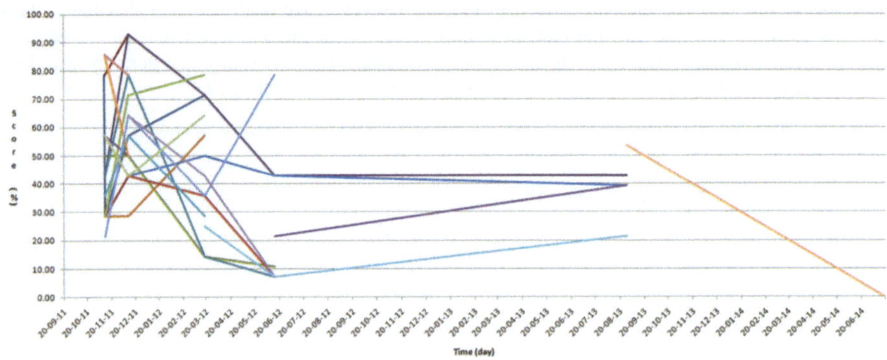

Fig. 14 Longitudinal score (IQ) of 18 individuals with more than two recordings. Scores on IQ for 18 individuals having more than two recordings and nonzero score are shown in Fig. 14. The scores are of higher elevation compared to that for total score of previous picture. Upward trend of scores is more prominent in the second interview, compared to the previous figure

Appendix

To study the deviation of the observed growth from nonparametric response curve derived under mild assumption and to have a bound on maximum fluctuation of deviations, consider these errors to be martingale differences. An assignable cause may be looked into, if the fluctuations are of different type other than specified by

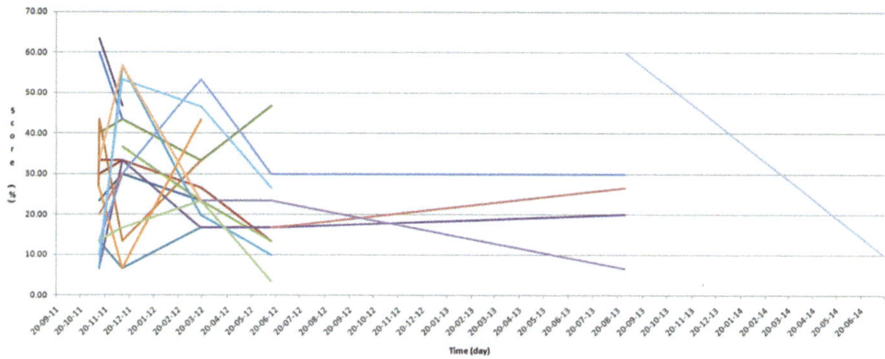

Fig. 15 Longitudinal score (Math) of 19 individuals with more than two recordings. Scores on Math for 19 individuals with more than two recordings are shown in Fig. 15. The scores are of lower elevation compared to those for total score and IQ scores of previous pictures. Upward trend of scores is prominent in general immediately after the first interview

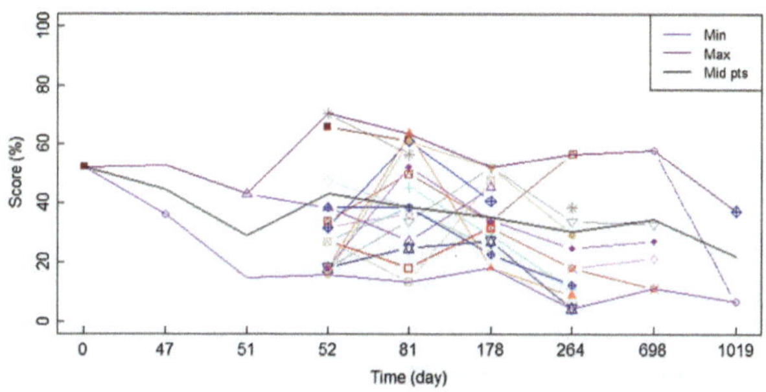

Fig. 16 Band of growth curves (Math + IQ) of 19 Individuals with >2 recordings. The band containing 19 longitudinal curves on total score is shown. The curve joining mid points of the upper and lower band is also shown in Fig. 16. The curve in the middle represents an estimate of overall growth in total scores under the assumption that errors are symmetric. With less number of curves, the picture is less messy. An individual scoring zero in IQ, scored a positive mark in Math. Thus, the number of individuals with nonzero total score is 19 here

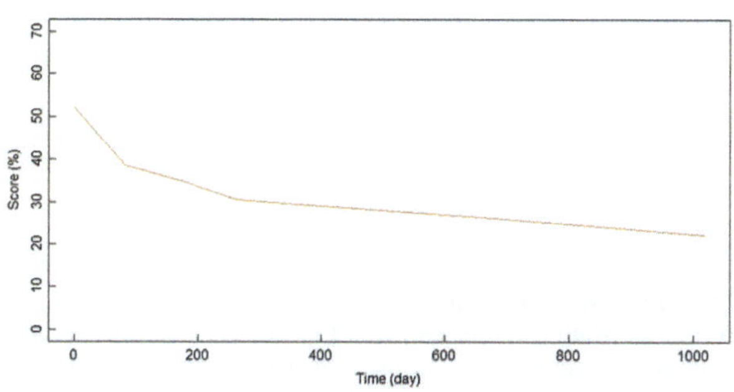

Fig. 17 Lowess on mid points of growth band (Math + IQ) of 19 individuals with >2 recordings. A nonparametric growth curve of total score by lowess regression on mid points of the band with $f = 2/3$ is shown in Fig. 17. The curve goes down over time towards stability

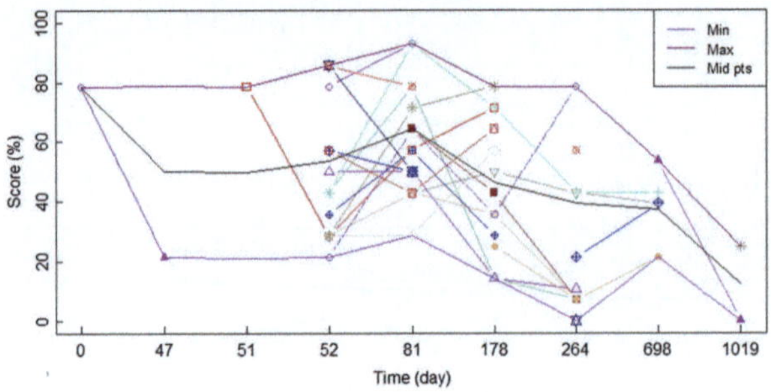

Fig. 18 Band of growth curves (IQ) of 19 individuals with >2 recordings. A similar band containing of 18 longitudinal curves on IQ score is shown in Fig. 18. Here again, the curve in the middle represents an estimate of overall growth of IQ score when errors have symmetric distribution. The pattern of growth is clearer with less number of individuals

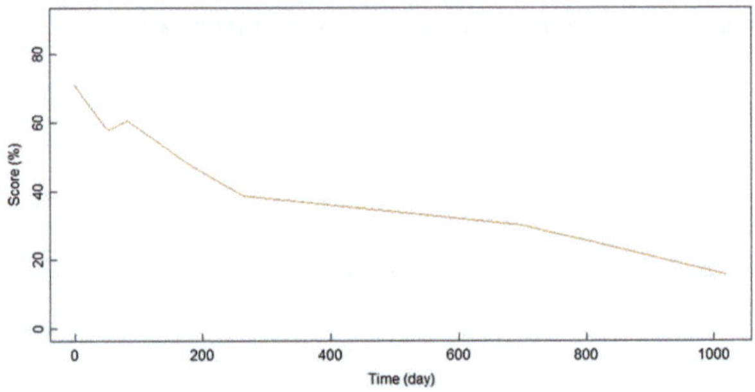

Fig. 19 Lowess on mid points of growth band (IQ) of 18 Individuals with >2 recordings. A nonparametric growth curve of IQ score by lowess regression on midpoints of the band with $f = 2/3$ is shown in Fig. 19. The curve has slight upward trend initially after a drop, and gradually lowers over time. This feature of the graph may be due to the assumption made on symmetric error

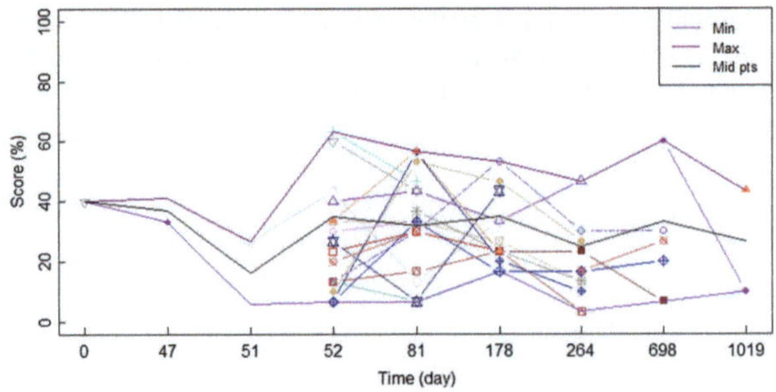

Fig. 20 Band of growth curves (Math) of 19 individuals with >2 recordings. The band containing 19 longitudinal curves on Math score is shown in Fig. 20. The curve in the middle represents an estimate of overall growth of Math score. The oscillatory pattern of growth has slight upward trend. This becomes clear with less number of individuals

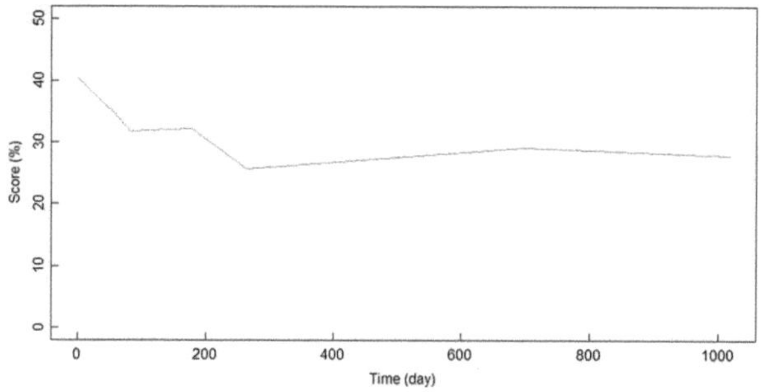

Fig. 21 Lowess on midpoints of growth band (Math) of 19 individuals with >2 recordings. A nonparametric growth curve of Math score by lowess regression on midpoints of the band with $f = 2/3$ is shown in Fig. 21. The curve indicates an upward trend, after slowing down slightly at the initial stage

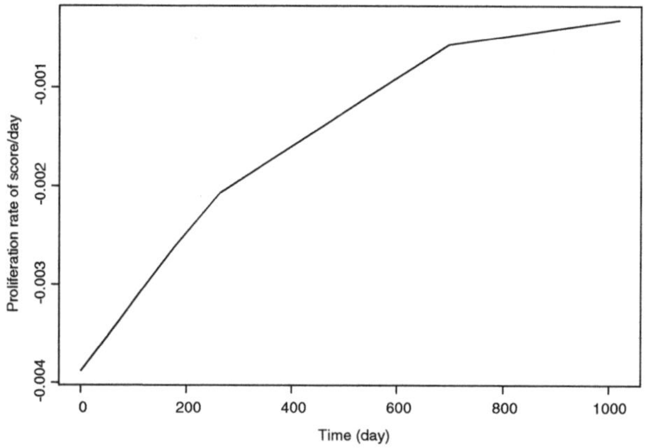

Fig. 22 Proliferation rate of 19 persons total score: wt. exp$(-x)$; spline. We compute the proliferation rate based on the growth curve shown in Fig. 17. Proliferation rate $\frac{d}{dt}\log y$ over 9 time points for total score y is computed based on a technique proposed in Dasgupta (2015), with exponentially decaying weights attached to empirical slopes computed from data pairs at different time points with respect to a fixed time point t of interest. Weighted mean of the empirical slopes is considered at derivative stage of computation and the package smooth.spline with spar $= 0.01$ at smoothing stage in SPlus provides the proliferation rate at the time point t. The curve in Fig. 22 has a sharp rise in the beginning and this gradually slows down towards the end. With less number of 19 individuals, the curve is different compared to that with 47 individuals shown in Fig. 10

martingale difference. We may obtain a bound on extreme fluctuation of observations from response curve from the assumption that the errors are martingale differences.

The following moment bound of Dasgupta (1993) holds for general stochastic processes. This includes martingales as a special case. Below we provide a slightly modified version of the result.

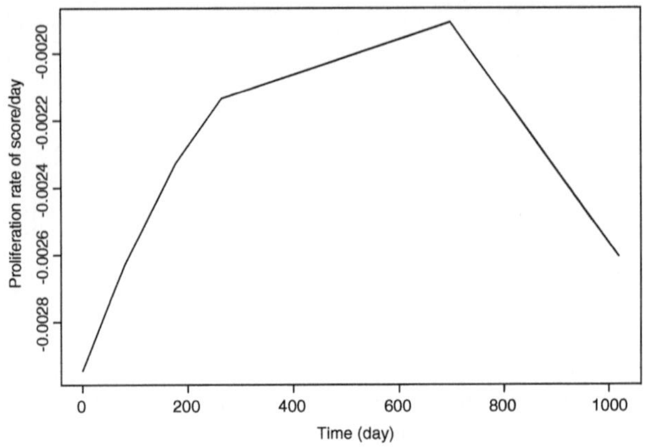

Fig. 23 Proliferation rate of 18 persons IQ score: wt. exp($-x$); spline. Proliferation rate $\frac{d}{dt} \log y$ over 9 time points for IQ score y is computed like Fig. 10, with exponentially decaying weights attached to empirical slopes computed from data pairs at different time points with respect to a fixed time point t of interest in Fig. 23. Weighted mean of the empirical slopes is considered at derivative stage of computation and the package smooth.spline with spar $= 0.01$ at smoothing stage in SPlus provides the proliferation rate at the time point t. A sharp rise of proliferation rate is seen in the beginning, which gradually slows down and then a downward trend of proliferation rate is seen. The rate of fall is sharper towards the end

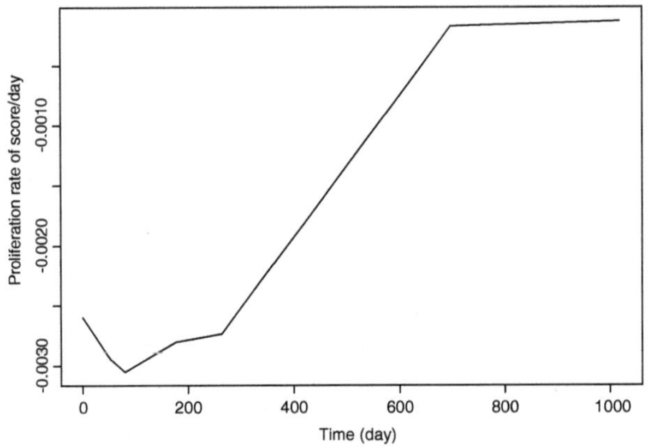

Fig. 24 Proliferation rate of 19 persons Math score: wt. exp($-x$); spline. Proliferation rate $\frac{d}{dt} \log y$ over 9 time points for Math score y is computed based on a technique proposed in Dasgupta (2015), with exponentially decaying weights attached to empirical slopes computed from data pairs at different time points with respect to a fixed time point t of interest in Fig. 24. Weighted mean of the empirical slopes is considered at derivative stage of computation and the package smooth.spline with spar $= 0.0001$ at smoothing stage in SPlus provides the proliferation rate at the time point t. A little bit downward trend is seen in the beginning reaching a minimum near 80 days, then the curve gradually moves up fast to reach a high value at time $t = 700$ days. The curve reaches stability at the far end. Rise of proliferation rate is slow towards the end

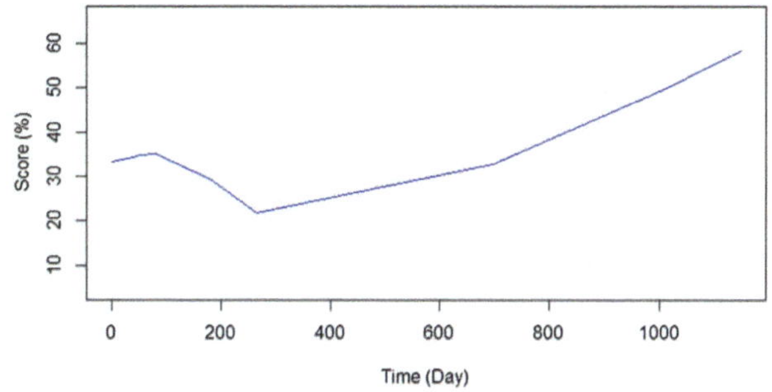

Fig. 25 Lowess on median of growth band (Math + IQ) of 47 individuals. We now examine the data with assumption of symmetric error dropped. Instead, we consider median to calculate the response curve. A nonparametric growth curve of total score by lowess regression on a robust estimate of central tendency as median scores at time points fixed on dates of interview in the band, with $f = 2/3$ is shown in Fig. 25. The curve dispenses off the stringent assumption that error components are symmetric. The curve has an initial upward trend for a while, then it goes down and reaches a minimum when initial boost from interaction with interviewer erodes; then the curve gradually shows an upward trend over time

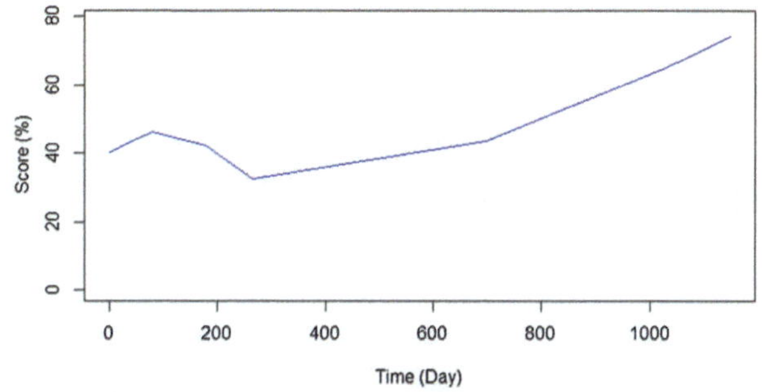

Fig. 26 Lowess on median of growth band (IQ) of 47 individuals. Growth curve of IQ score by lowess regression on median of scores at time points fixed on dates of interview in the band, with $f = 2/3$ is shown. Elevation of this curve is a bit higher than the previous graph on total score. Like in the previous curve, this has initial upward trend for a while, then it goes down and reaches a minimum when initial boost from interaction with interviewer possibly erodes; then the curve gradually shows an upward trend over time. The pattern of the curve in Fig. 26 is not similar to that shown in Fig. 7, indicating that the assumption of symmetric error component for IQ scores may not hold

Theorem A *Let* $\{X_i, i \geq 1\}$ *be a stochastic process with* $E[sgn(S_{i-1})X_i|\ |S_{i-1}|] \leq 0$, $E(\sum_{i=1}^{n} \pm X_i)^2 \leq n\beta^*_{2,n}$, *where* $S_i = \sum_{j=1}^{i} X_j$, $\gamma_{\nu,n} = E|X_n|^\nu$, $\beta^*_{\nu,n} = \max_{1\leq j\leq n} \gamma_{\nu,j}$. *If the l.h.s. of (1) is finite, then for* $\nu \geq 2$

$$E|S_n|^\nu \leq c_\nu n^{\nu/2}\beta^*_{\nu,n}, \text{ where } c_\nu = [2(\nu-1)\delta]^{\nu/2} \tag{1}$$

and for large n, $\delta \approx (1 + \frac{\nu}{2n})$.

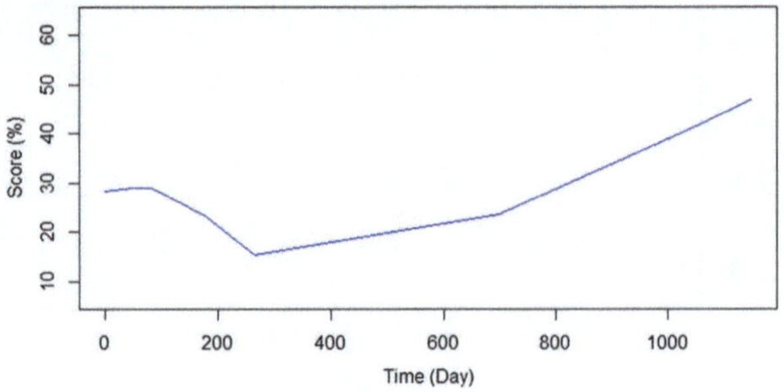

Fig. 27 Lowess on median of growth band (Math) of 47 individuals. Growth curve of Math score by lowess regression on median of scores at time points fixed as dates of interview in the band, with $f = 2/3$ is shown in Fig. 27. Elevation of this curve is lower than the previous two graphs. Like the previous curve, this has initial upward trend for a while. Then it goes down further compared to Fig. 9 and reaches a minimum when initial boost from interaction with interviewer possibly erodes; then the curve gradually shows an upward trend over time. The pattern is similar to that for Fig. 9

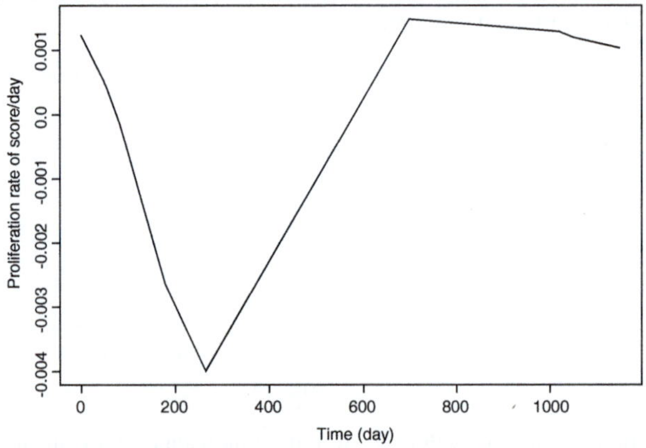

Fig. 28 Proliferation rate of 47 persons total score (med): wt. exp$(-x)$; spline. Proliferation rate $\frac{d}{dt}\log y$ over 11 time points for total score y is computed like Fig. 10, with exponentially decaying weights attached to empirical slopes computed from data pairs at different time points with respect to a fixed time point t of interest in Fig. 28. Weighted mean of the empirical slopes is considered at derivative stage of computation and the package smooth.spline with spar $= 0.0001$ at smoothing stage in SPlus provides the proliferation rate at the time point t. A downward trend in the beginning is seen and then a typical sharp rise is seen to reach a maximum at $t = 700$ days, the peak value slightly crossing the initial value. A bit downward trend is seen towards the end to reach stability of proliferation rate

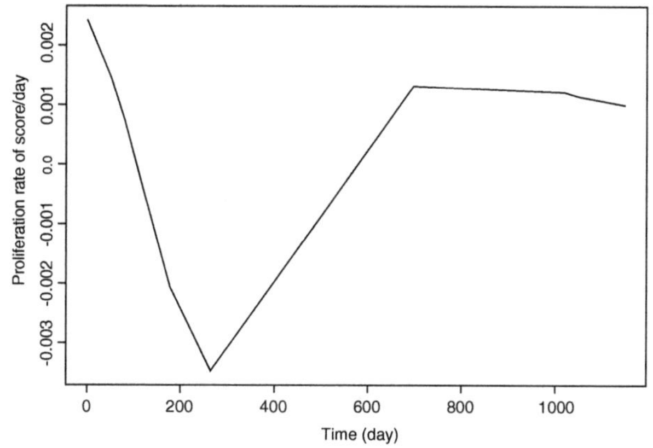

Fig. 29 Proliferation rate of 47 persons IQ score (med): wt. $\exp(-x)$; spline. Proliferation rate $\frac{d}{dt}\log y$ over 11 time points for IQ score y is computed based on a technique proposed in Dasgupta (2015), with exponentially decaying weights attached to empirical slopes computed from data pairs at different time points with respect to a fixed time point t of interest in Fig. 29. Weighted mean of the empirical slopes is considered at derivative stage of computation and the package smooth.spline with spar $= 0.0001$ at smoothing stage in SPlus provide the proliferation rate at the time point t. A steep fall in the beginning is seen to reach a minimum at $t = 250$ days and then a rise upto $t = 700$ days is seen with peak value less than the initial value of proliferation rate seen at start. Finally stability at the far end is seen

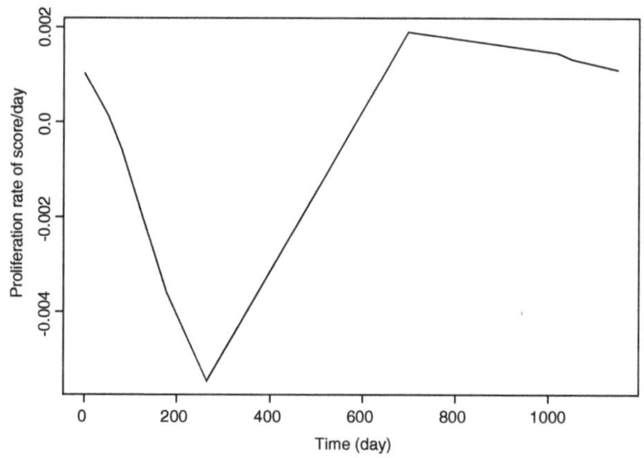

Fig. 30 Proliferation rate of 47 persons Math score (med): wt. $\exp(-x)$; spline. Proliferation rate $\frac{d}{dt}\log y$ over 11 time points for Math score y is computed like Fig. 10, with exponentially decaying weights attached to empirical slopes computed from data pairs at different time points with respect to a fixed time point t of interest in Fig. 30. Weighted mean of the empirical slopes is considered at derivative stage of computation and the package smooth.spline with spar $= 0.0001$ at smoothing stage in SPlus provides the proliferation rate at the time point t. A sharp fall in the beginning is seen reaching a minimum at $t = 250$ days and then a sharp rise of the curve is seen, with rate approximately 0.002 at $t = 700$ days, to cross the initial value of rate. Ability to learn mathematics is commendable in individuals. Finally, the proliferation rate shows a downward trend to reach stability at the far end

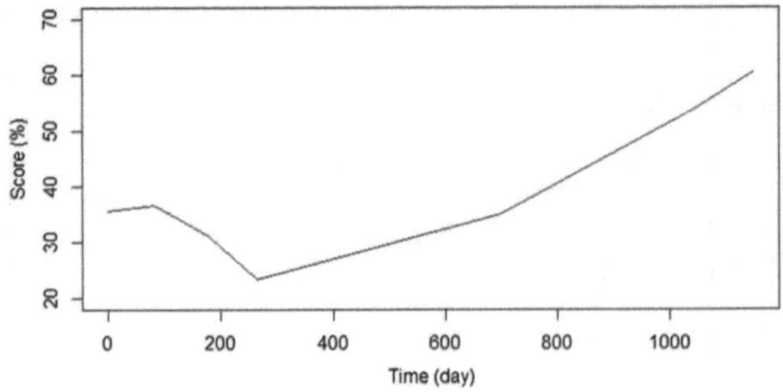

Fig. 31 Lowess on mean of growth curves (Math + IQ) of 47 individuals. Although not robust, mean is a natural estimate in sample for central tendency. Instead of median, we consider lowess with $f = 2/3$ on mean of growth curves in Fig. 31. Rise of the curve for total score is seen at the beginning; it then reaches a minimum and gradually rises again till the end

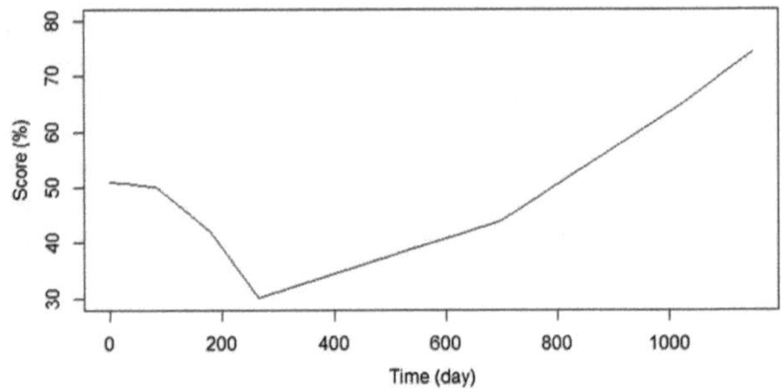

Fig. 32 Lowess on mean of growth curves (IQ) of 47 individuals. In this lowess curve with $f = 2/3$, the fall from high initial value is steep. Then the curve in Fig. 32 rises again towards a high value of 66.95%

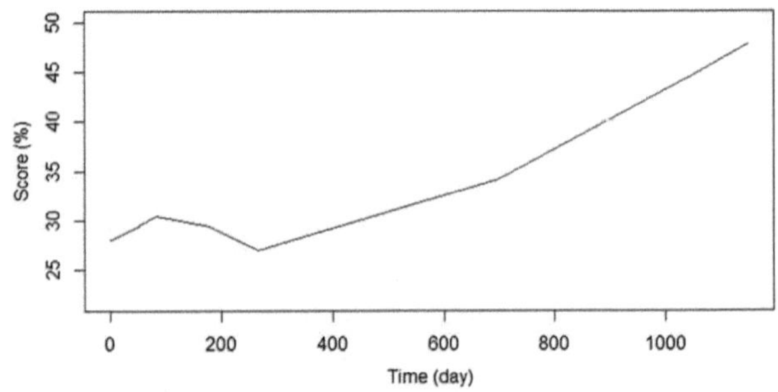

Fig. 33 Lowess on mean of growth curves (Math) of 47 individuals. In Fig. 33, unlike Fig. 32 the fall is not that steep in this lowess curve with $f = 2/3$ for Math score. After a fall, the curve rises again to reach a value of 47.80%. This is lower than the corresponding value 66.95% in IQ, shown in the previous figure

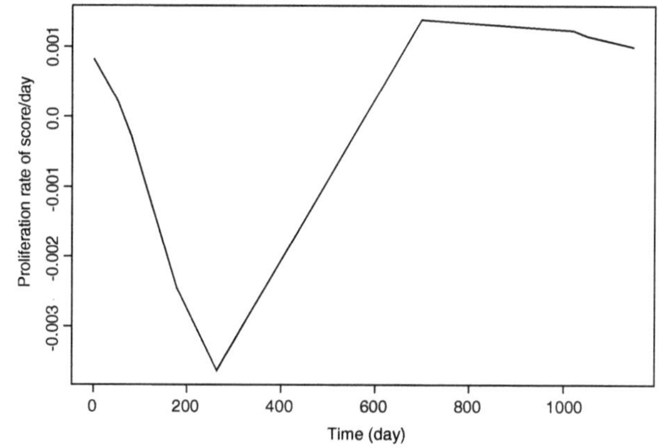

Fig. 34 Proliferation rate of 47 persons total score (mean): wt. exp($-x$); spline. The proliferation rate curve in Fig. 34 with spar = 0.0001 mimics the pattern of Fig. 28 where median is considered in place of mean. The patterns are very similar in either case of mean or median being used to estimate the growth curve of total score, the similarity of Figs. 27 and 31 is reflected in the similarity of proliferation rate curves computed under different assumptions

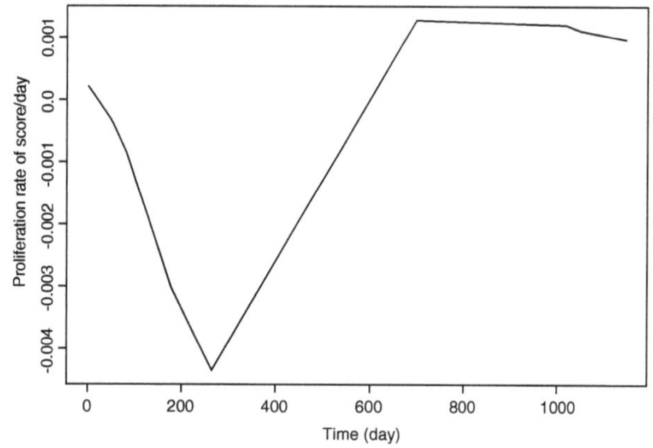

Fig. 35 Proliferation rate of 47 persons IQ score (mean): wt. exp($-x$); spline. Proliferation rate with spar = 0.0001 in Fig. 35 with mean values for IQ rises to a higher level than the initial proliferation rate at start, unlike that of Fig. 29 with median of IQ scores where initial value of the rate is the highest

An extra term 2 in c_ν above appears due to the fact that expectation of maximum of the terms in (2.2) of Dasgupta (1993) is bounded above by sum of the expectations in (2.3) therein, leading to an extra factor 2; i.e., the correct expression is $E\max(|S_n|^{\nu-2}X_n^2, |S_n^*|^{\nu-2}X_n^2) < E(|S_n|^{\nu-2}X_n^2 + |S_n^*|^{\nu-2}X_n^2)$. The bound is useful in estimating remainder from the main part in nonlinear statistics that arise in many situations related to limit theorems, see, e.g., Dasgupta (1994). The modification does not affect the results of Dasgupta (1994) as L is a generic positive constant therein.

Consider $M_i = \sum_{j=1}^{i} y_j$, where M is a martingale and y s are the martingale differences. For a martingale M with finite νth moment, one may write From Doob's inequality

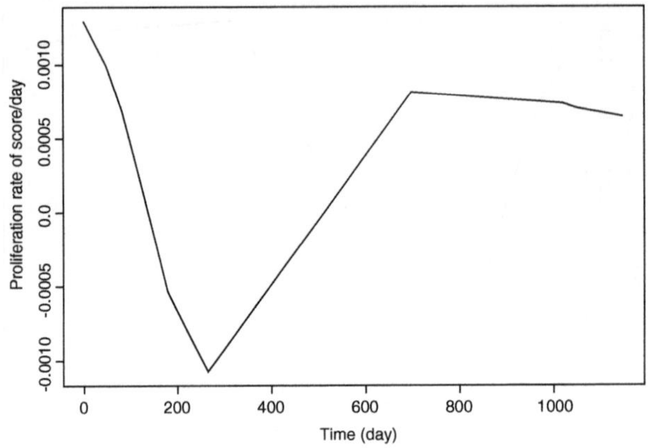

Fig. 36 Proliferation rate of 47 persons math score (mean): wt. exp($-x$); spline. Proliferation rate with spar = 0.0001 in Fig. 36 with mean values for Math scores has a sharp downward trend in the beginning. Then the curve rises again to achieve a peak with lower value than the initial value at start, before stability is achieved; this pattern is unlike Fig. 30 with median score, where the peak is higher than the initial value

$$P(\max_{1\leq i\leq n} |M_i| \geq c) \leq c^{-\nu} E|M_n|^{\nu} \tag{2}$$

Thus, if a finite $\nu(\geq 2)$ th order moment of martingale differences y_j exist, i.e.,

$$\sup_{j\geq 1} E|y_j|^{\nu} < \infty$$

then from (1) and (2) one may write

$$P(\max_{1\leq i\leq n} |M_i| \geq \lambda n^{1/2}) \leq \lambda^{-\nu} c_{\nu} \beta^*_{\nu,n} \tag{3}$$

Next consider the case where all the moments of martingale differences exist, but the moment generating function need not necessarily exist. Consider the following moment bounds for the martingale differences y_j.
Type 1:

$$\sup_{j\geq 1} E|y_j|^{\nu} \leq L e^{w_0 \nu^{\alpha}} \tag{4}$$

$\forall \nu > 1$, and for some $L > 1$, where $w_0 > 0$, $\alpha > 1$. In what follows $L(> 1)$ is a generic constant. The above condition is equivalent to

$$\sup_{j\geq 1} E\exp[s\{\log_e(1+|y_j|)\}^{\alpha/(\alpha-1)}] < \infty \tag{5}$$

where $s = w_0^{-1/(\alpha-1)}$, see (4.15)–(4.16) and Appendix in p 87 of Dasgupta (2013) for similar assumptions. Assumption (4) is equivalent to finiteness of m.g.f. in a neighborhood of zero compared to that in Dasgupta (2015) for the transformed random variable $\{\log_e(1+|y_j|)\}^{\alpha/(\alpha-1)}$. The original variable y_j has moment bounds of high magnitude. After logarithmic transformation, the variables are tamed, and some power of the transformed random variable y_j possess m.g.f.

Probability bound given in (3) may be sharpened under the assumption (4)/(5) by minimizing the r.h.s. of (3) with respect to ν under specific moment bound. Under (4), write

$$P(\max_{1\leq i\leq n}|M_i| \geq \lambda n^{1/2}) \leq \lambda^{-\nu} c_\nu \beta^*_{\nu,n} < Le^{w\nu^\alpha}\lambda^{-\nu} = P^*(\text{say}),\ w > w_0 \tag{6}$$

Next write $\log P^* = \log L + w\nu^\alpha - \nu\log\lambda$. At the minimum value $\frac{d}{d\nu}\log P^* = 0$ provides, $\alpha w\nu^{\alpha-1} - \log\lambda = 0$, with $\nu = (\frac{\log\lambda}{\alpha w})^{1/(\alpha-1)}$. Write $\log P^* = \log L + (\alpha w\nu^\alpha - \nu\log\lambda) + (1-\alpha)w\nu^\alpha = \log L + (1-\alpha)w\nu^\alpha = \log L + (1-\alpha)$ $w(\frac{\log\lambda}{\alpha w})^{\alpha/(\alpha-1)}$. The second derivative $\frac{d^2}{d\nu^2}\log P^* = \alpha(\alpha-1)w\nu^{\alpha-2} > 0$. Hence, we have the following result.

Theorem 1 *Under the moment bound given in (4)/(5) for martingale differences, the following excessive deviation result holds for the martingale:* $\{M_i : 1\leq i\leq n\}$

$$P(\max_{1\leq i\leq n}|M_i| \geq \lambda n^{1/2}) \leq Le^{-(\alpha-1)w(\frac{\log\lambda}{\alpha w})^{\alpha/(\alpha-1)}},\ \lambda\to\infty \tag{7}$$

where $w > w_0$ *is given in (4).*

Remark 1 The above result is sharper than the bound given in (3). Since $\alpha > 1$, the decaying order of λ as seen from $\lambda^{-(\alpha-1)w(\frac{\log\lambda}{\alpha w})^{1/(\alpha-1)}}$, is greater than any polynomial power, when λ is large; $\lambda\to\infty$.

Next we consider the following type of moment bounds for martingale differences: *Type 2*:

$$\sup_{j\geq 1} E|y_j|^\nu \leq L^\nu e^{\alpha\nu\log\nu} \tag{8}$$

$\forall\nu\geq 2$, where $L > 0,\ \alpha > 1$. The above condition is implied by

$$\sup_{j\geq 1} E\exp(s|y_j|^{1/\alpha}) < \infty \tag{9}$$

where $0 < s < s_0 = \alpha e^{-1}L^{-1/\alpha}$, see (4.34)–(4.35) of Dasgupta (2013) for similar assumptions. Assumption (12) is equivalent to finiteness of m.g.f. in a neighborhood of zero for the transformed random variable $|y_j|^{1/\alpha}$. The assumption is weaker than existence of m.g.f. as $\alpha > 1$.

Sharper results on excessive deviation may be computed in this setup. Under (8), write

$$P(\max_{1\le i\le n} |M_i| \ge \lambda n^{1/2}) \le \lambda^{-\nu} c_\nu \beta^*_{\nu,n} \le \lambda^{-\nu} L^\nu e^{(\alpha+r)\nu\log\nu} = P^*(\text{say}) \qquad (10)$$

where $r = \frac{1}{2}$ for $\nu \ll n$, and $r = 1$ for unrestricted ν.

Next write $\log P^* = \nu \log L + (\alpha + r)\nu \log \nu - \nu \log \lambda$. At the minimum value $\frac{d}{d\nu} \log P^* = 0$ provides, $\log L + (\alpha + r) \log \nu + (\alpha + r) - \log \lambda = 0$, with $\log \nu = (\log \lambda - \log L)/(\alpha + r) - 1$. Then $\log P^* = \nu(\log L + (\alpha + r) \log \nu + (\alpha + r) - \log \lambda) - \nu(\alpha + r) = -\nu(\alpha + r)$. Thus $P^* = e^{-(\alpha+r)e^{(\log\lambda-\log L)/(\alpha+r)-1}}$, where $r = \frac{1}{2}$, if $n \gg \nu = O(e^{\log\lambda/(\alpha+\frac{1}{2})})$, i.e., $\lambda = O(n^{\alpha+\frac{1}{2}})$; and $r = 1$, otherwise, i.e., for deviations λ of higher order.

The second derivative $\frac{d^2}{d\nu^2} \log P^* = (\alpha + r)/\nu > 0$. Hence, we have the following.

Theorem 2 *Under the moment bound given in (8)/(9) with* $\alpha > 1$, *the following excessive deviation result holds for the martingale* $\{M_i : 1 \le i \le n\}$*:*

$$P(\max_{1\le i\le n} |M_i| \ge \lambda n^{1/2}) \le e^{-(\alpha+r)e^{(\log\lambda-\log L)/(\alpha+r)-1}} \quad \lambda \to \infty \qquad (11)$$

where $r = \frac{1}{2}$, *if the deviation* $\lambda = O(n^{\alpha+\frac{1}{2}})$; *and* $r = 1$, *otherwise, i.e., for deviations of higher order,* $\lambda \ggg n^{\alpha+\frac{1}{2}}$.

We further consider the following type of bound, which is more stringent than the bounds considered above.

Type 3: Bound of Type 2 with a different parametric zone, where $\alpha \in (0, 1]$. *That is,*

$$\sup_{j\ge1} E \exp(s|y_j|^{1/\alpha}) < \infty, \ \alpha \in (0, 1] \qquad (12)$$

This ensures m.g.f. of y_j *exist, but* y_j $j \ge 1$ *may not be bounded. Nonuniform Berry-Esseen bound in CLT for such cases was considered in Dasgupta (2006) for independent random variables in a triangular array.*

For the moment bound of Type 3, take $\alpha \in (0, 1]$ *in (9). This assumption ensures the existence of m.g.f for the martingale differences* y_j, *however,* y_j *in martigale* M_n *may not be bounded as* $n \to \infty$. *We then have*

$$\sup_{j\ge1} E|y_j|^\nu \le L^\nu e^{\alpha\nu\log\nu}, \ \forall\nu \ge 2, L > 0, \ \alpha \in (0, 1] \qquad (13)$$

The above condition is implied by

$$\sup_{j\geq 1} E\exp(s|y_j|^{1/\alpha}) < \infty \tag{14}$$

where $0 < s < s_0 = \alpha e^{-1}L^{-1/\alpha}, \alpha \in (0, 1]$.

Calculations for the moment bound of Type 3 with $\alpha \in (0, 1]$ *is similar to that of Type 2 moment bound considered above. Proceeding in a similar fashion we obtain the following, which includes the case of bounded martingale differences in the limiting case when* $\alpha \to 0$.

Theorem 3 *Under the moment bound of Type 3, i.e., (8)/(9) with* $\alpha \in (0, 1]$, *the following excessive deviation result holds for the martingale* $\{M_i : 1 \leq i \leq n\}$*:*

$$P(\max_{1\leq i\leq n} |M_i| \geq \lambda n^{1/2}) \leq e^{-(\alpha+r)e^{(\log\lambda-\log L)/(\alpha+r)-1}}, \lambda \to \infty \tag{14}$$

where $r = \frac{1}{2}$, *if the deviation* $\lambda = O(n^{\alpha+\frac{1}{2}})$; *and* $r = 1$, *otherwise, i.e., for deviations of higher order,* $\lambda \ggg n^{\alpha+\frac{1}{2}}$.

Remark 2 The bound obtained above for large deviations on martingale is exponentially decaying, and sometimes even higher than that. For standardized sum of iid random variables CLT holds and the tail probability of normal distribution is $\Phi(-\lambda) \sim \frac{1}{\sqrt{2\pi}}\lambda^{-1}e^{-\lambda^2/2}, \lambda \to \infty$. The bound in (14) with $\alpha \to 0,\ r = 1/2$ is $O(e^{-\frac{\lambda^2}{2}(1+o(1))});\ \lambda \to \infty$.

Remark 3 Under the assumption that the scores are bounded variable, the observed residuals, i.e., deviations of data points from growth curve have a probabilistic bound specified by normal distribution, vide Remark 2. Any high departures of observations from the response curve, not in concordance with normal deviate, are to be assessed for assignable causes.

References

Dasgupta, R. (2015). Rates of convergence in CLT for two sample U-statistics in Non iid case and multiphasic growth curve. In R. Dasgupta (Ed.), *Growth curve and structural equation modeling* (Vol. 132, pp. 35–58). Springer Proceedings in Mathematics & Statistics.

Dasgupta, R. (1993). Moment bounds for some stochastic processes. *Sankhyā A*, *55*, 152–180.

Dasgupta, R. (1994). Nonuniform speed of convergence to normality while sampling from finite population. *Sankhyā A*, *56*(2 special issue), 227–337.

Dasgupta, R. (2006). Nonuniform rates of convergence to normality. *Sankhyā A*, *68*, 620–635.

Dasgupta, R. (2013). Non uniform rates of convergence to normality for two sample U-statistics in Non iid case with applications. In *Advances in Growth Curve Models: Topics from the Indian Statistical Institute* (Vol. 46, pp 60–88). Springer Proceedings in Mathematics & Statistics.

Duminy, P. A., Dreyar, H.J., Steyn, P. D. G., Behr, A. L., & Vos A.J. (1991). *Education for the student teacher 2*. South Africa: Maskew Miller Longman.

http://dspace.vidyanidhi.org.in:8080/dspace/bitstream/2009/4402/10/OSM-1993-041-9.pdf.

http://planningcommission.nic.in/reports/publications/pub_neregion.pdf.

http://timesofindia.indiatimes.com/india/7-of-8-Northeast-states-lag-behind-average-India-income/articleshow/5187537.cms.

Rider, P. R. (1957). The midrange of a sample as an estimator of the population midrange. *Journal of the American Statistical Association*, *52*(280), 537–542.

Distribution of Mangroves and Soil Parameters in the Lothian Island of Sundarbans, India—A GIS Approach

Hema Gupta (Joshi) and Monoranjan Ghose

Abstract Tropical mangrove ecosystems are highly productive and provide extensive ecosystem services. This paper analyzes distribution of 21 mangrove species (thirteen true mangroves and eight mangrove associates) in relation to various tidal and edaphic factors. Species density, soil parameters and tidal inundation were explored in 40 sites of the Lothian Island of the Western Sundarbans. Only five species and two soil parameters varied significantly under different tidal inundations. Single-linkage hierarchical classification of species densities generated seven distinct species clusters. Stepwise regression of species densities with soil parameters explained a variation of 42% or higher. Calcium, phosphorus, nitrogen, pH, salinity and sand explained the maximum variation in species densities. Statistical method 'kriging' in the Arc GIS generated maps predicting distribution of species densities and selected soil parameters over the island. Maps illustrated highly saline northern mudflat region with a distinct seaward fringe of *Avicennia alba* and herb *Acanthus ilicifolius*. Soil nutrients like nitrogen, phosphorus and cations like sodium, calcium and magnesium were concentrated in the northern half of the island. Most of the species were concentrated in the middle- and southern-ridged portion of the island having lower soil salinity.

Keywords Anova · Kriging · Mangrove · Stepwise regression

H. Gupta (Joshi) (✉) · M. Ghose
Agricultural and Ecological Research Unit, Indian Statistical Institute, 203 B.T. Road, Kolkata 700108, India
e-mail: hemagupta.gupta123@gmail.com

Present Address
H. Gupta (Joshi)
Department of Botany, Visva-Bharati, Bolpur, Santiniketan 731235, West Bengal, India

R. Dasgupta (ed.), *Advances in Growth Curve and Structural Equation Modeling*,
https://doi.org/10.1007/978-981-13-0980-9_2

1 Introduction

The term 'Mangrove' refers to a group of taxonomically heterogeneous woody shrubs and trees growing in intertidal zone of tropical and subtropical coasts. They have a restricted distribution between 31°N and 38°45′S (Hosakawa et al. 1977; Tomlinson 1986) due to the sensitivity of mangroves to frost and low temperature (Walter 1977). Recently, mangrove distribution is reported in temperate latitudes (Krauss et al. 2008). Worldwide there are 114 species of true mangroves belonging to 66 genera with species richness being greatest in the Indo-Pacific region (Tomlinson 1986). Mangroves provide income for human society as well as ecosystem services playing a crucial role in the maintenance of biodiversity, waste assimilation, cleansing, recycling and renewal as well as in protecting coastal areas from disturbance events (Alongi 1993; Dahdous-Guebas et al. 2005; Hussain and Badola 2010; Vo et al. 2012).

Mangrove communities often exhibit distinct patterns of species distribution (Chapman 1976; Lugo and Snedakar 1974). Any external force, substance or condition, which surrounds and affects the life of an organism in any way, becomes a factor of its environment and may be living (biotic) or non-living (abiotic). Several studies correlated salinity with the mangrove vegetation and productivity (Chen and Twilley 1998, 1999; Lugo 1980; Mall et al. 1987; Ukpong 1991). Local patterns of tidal inundation further influence soil characteristics that control species zonation of mangrove wetlands (Banerjee 1987; Naidoo 1980; Saha and Choudhury 1995; Walsh 1974; Watson 1928). Patches and zones of forest composition result from complex gradients of hydroperiod and soil conditions, such as nutrient limitation (Boto and Wellington 1984; Feller 1995; Ukpong 1998) and abiotic stressors as salinity and sulfide (Ball 1998; Mckee 1993; Nickerson and Thibodeau 1985). Lugo (1980) portrayed 35 interconnected factors responsible for species distribution. Other findings emphasized on joint actions of biological (propagule predation: MacGuiness 1997b; Smith 1987; competition: Smith 1988b) and physical processes (Lee 1999; Smith 1992) responsible for vegetation pattern. Nutrient availability in the mangrove soils varies between and within mangrove stands (Feller et al. 2003a) and is further controlled by a variety of biotic and abiotic factors such as tidal inundation, elevation in tidal frame, redox status, microbial activities of soil and plant activities (Reef et al. 2010). Soil pH significantly affects plant growth, primarily due to the change in availability of both essential elements such as phosphorus (P), as well as non-essential elements such as aluminium (Al) that can be toxic to plants at elevated concentrations (Black 1993; Slattery et al. 1999). The importance of both soil salinity and pH for the growth of mangroves has been emphasized by Wakushima et al. (1994a, 1994b). In an earlier study on limited sites, we also described forest structure and species distribution along soil salinity and pH gradient in the Sundarbans mangroves (Joshi and Ghose 2003).

Okimoto et al. (2008) utilized growth curve analysis method to study net carbon fixation of mangrove *Rhizophora apiculata* by drawing a growth curve using data of above- and below-ground biomass at 5, 10 and 15 years of age. They estimated net carbon fixation as 1.1 to 35.2 Mg C ha^{-1} $year^{-1}$. Nguyen et al. (2015) studied growth of the grey mangrove *Avicennia marina* in response to salinity by growing seedlings hydroponically in dilutions of seawater amended with nutrients. They analysed whole plant growth characteristics in relation to dry mass accumulation and its allocation to different parts, and concluded *Avicennia marina* to be an obligate halophyte that required saline conditions to develop transport systems needed to sustain water use and carbon gain. Nazim et al. (2013) determined age and radial growth in grey mangrove *Avicennia marina* by using two different methods—first, the dendrometer technique in which a dendrometer was fixed properly on the main tree trunk in order to determine the monthly radial growth, and second, by taking manual measurements from the healthy branches of the upper and the lower portions of the main trunk. The radial growth of main trunk as well as the branches significantly increased with the passage of time. The maximum growth rate (9.54 ± 0.95 cm) was in the month of July, whereas it was minimum (0.58 ± 0.058 cm) in the month of October. The growth rate was higher in the upper branches than the lower part of the main trunk. We plan to undertake a similar type of work on mangroves in the soil of Sundarbans.

India is one of the most mangrove-rich countries having 2.7% of global total area of mangrove (Giri et al. 2011). Sundarbans shelters one of the most important mangrove communities of the world. 37 obligate mangroves and 32 mangrove associates are reported from the entire Sundarbans (Banerjee et al. 1989; Naskar and Guhabakshi 1987). In this paper, we report on the distribution of mangrove species in relation to tidal inundation and some physicochemical parameters of the substratum.

2 Materials and Methods

2.1 Study Area

Sundarbans is situated in the delta of the rivers Ganges and Brahmaputra of the Indian subcontinent. The total area of mangrove forest in the Indian Sundarbans is estimated to be 4,267 km^2 including rivers, creeks and canals (Naskar 2004). Earlier, Forest Survey of India (1999) estimated the area of Indian Sundarbans as 2125 km^2 based on satellite imagery, excluding the network of creeks and backwaters, which are part and parcel of this ecosystem. The present study was undertaken at Lothian Island, situated in Saptamukhi estuary complex of the Sundarbans Biosphere Reserve from November 1999 to February 2003. It is a small island (~38 km^2), which extends from 88°18′10″E to 88°21′30″E longitude and 21°32′50″N to 21°42′30″N latitude. A North–South spinal road approaches from northern coast up to nearly middle of the island. Diurnal tides inundate the northern half of the island, while middle and southern portions inundate during spring tides (Plates 1 and 2). There are some areas in the interior of the island where occasional storms and rains are the only source of water.

Plate 1 *Avicennia alba* zone

Plate 2 Mangroves along the creek of Sundarbans

2.2 *Vegetation and Soil Sampling*

Vegetation and soil measurements were obtained in 40 sites distributed randomly and covering a total area of 3562 m^2 (Fig. 1). At each site quadrat of 4 m × 16 m (Joshi and Ghose 2002) was laid for sampling trees including shrubs and climbers. Herbs were sampled by laying four 1 m × 1 m quadrat nested within the 4 m × 16 m quadrat. In each quadrat, all the trees (plants >1 cm dbh) were identified and their number and diameter at breast height were recorded. The category of tidal inundation (diurnal tides, spring tides, summer spring tides and no tides) was also recorded. Density (N ha^{-1}) was calculated for each species in the 40 sites according to Cintron and Schaeffer-Novelli (1984).

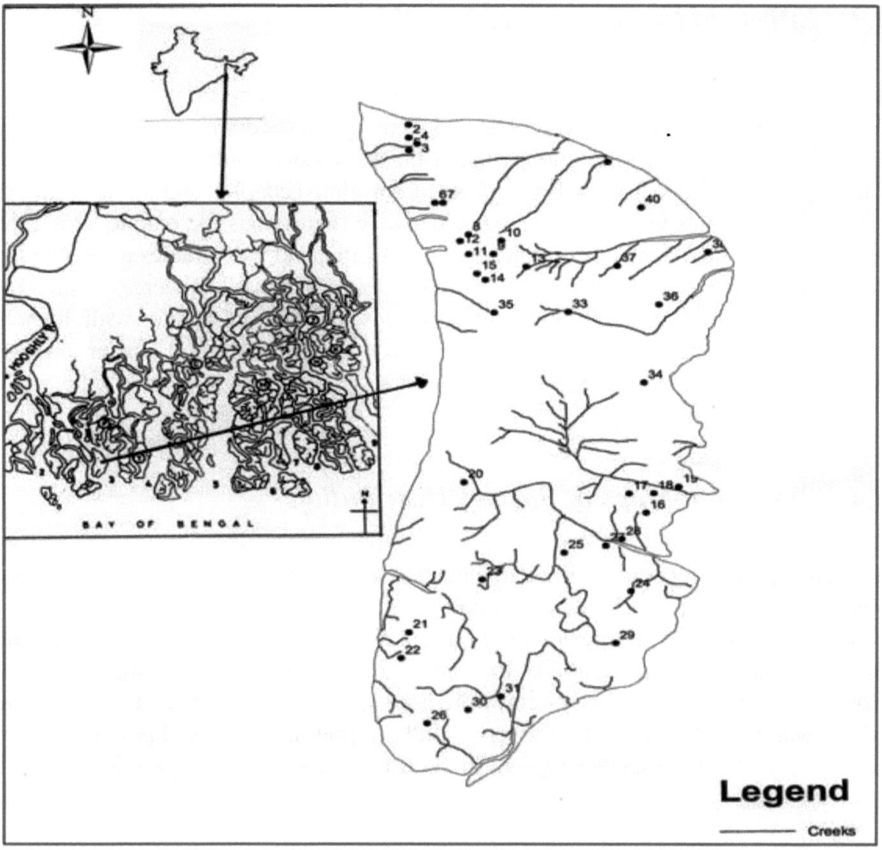

Fig. 1 Boundary map of Lothian Island showing the 40 study sites (Gupta Joshi and Ghose 2014)

Three soil samples were collected from each site from a rooting depth of 15 cm. The samples were air-dried, crushed using a pestle and mortar, and then passed through a 10-mesh (2 mm) screen before analysis. Standard methodology as described by Jackson (1973) was followed for estimating soil salinity, pH, available potassium and sodium. The International pipette method was employed for particle size analysis (Piper 1960). Organic carbon was estimated by the wet digestion method (Walkley and Black 1934). Available nitrogen was estimated following the potassium permanganate oxidation method (Subiah and Asiza 1956). Olsen's bicarbonate extraction method (Olsen et al. 1954) was employed for the estimation of available phosphorus. Available calcium and magnesium were estimated using EDTA or Versenate titration method (Baruah and Barthakur 1997).

2.3 Effects of Environmental Variables

The impact of tidal inundation on species densities was determined with one-way ANOVA. A single-linkage hierarchical cluster analysis was done on species densities using absolute value of correlation as the distance measure resulting in segregation of a few species clusters. To assess the effect of correlated edaphic and biotic variables on species densities in a species cluster, stepwise regressions were done taking the soil parameters and densities of other species as independent variables to predict the density of a species in that cluster. Stepwise regression equations were prepared for the species that were present in at least 10% sites, i.e. 4 out of 40 sites with the help of statistical package BMDP 7.01 (module 2R). The ANOVA and cluster analysis were done in the computer package SPSS 7.5.1.

2.4 Prediction Maps of Species Distribution

The maps predicting some soil parameters and species densities over Lothian Island were prepared using a statistical method 'kriging' in the Arc GIS 8 software. For generating the layers of species distribution, the species densities at 40 sites and the exact geographical location of the sites (latitude and longitude values obtained from the Global Positioning System) were used as the attributes. The map of Lothian Island was digitized from the Survey of India topographic sheet. Prediction maps were prepared for the species present in 10 or more sites.

3 Results

3.1 Species Composition

A total of 21 species (12 trees, 3 shrubs, 3 herbs and 3 climbers), including 13 true mangroves and 8 mangrove associates, were reported from 40 sites (Table 1). Five species—*Acanthus ilicifolius*, *Aegialitis rotundifolia*, *Aegiceras corniculatum*, *Avicennia marina* and *Ceriops decandra*—were present at more than 50% sites. Four species—*Brownlowia tersa*, *Bruguiera gymnorrhiza*, *Ceriops tagal* and *Lumnitzera racemosa*—occurred in single site each. A detailed analysis of vegetation and soils of these 40 sites is reported in an earlier study (Joshi and Ghose 2014). In the following account, effects of tidal inundation, edaphic and biotic factors on mangrove species distribution are given.

Table 1 Species encountered and their percentage distribution within 40 sites of Lothian Island (Gupta Joshi and Ghose 2014)

Species	Abbreviation	Family	% of sites
Trees			
Aegiceras corniculatum (L.) Blanco	AC	Myrsinaceae	50.0
Avicennia alba Blume	AA	Avicenniaceae	20.0
Avicennia marina (frosk.) Vierh	AM	Avicenniaceae	75.0
Avicennia officinalis L	AO	Avicenniaceae	32.5
[a]*Brownlowia tersa* (Linn.) Kostern	BT	Tiliaceae	2.5
Bruguiera gymnorrhiza (L.) Lamak	BG	Rhizophoraceae	2.5
Ceriops decandra (Griff.) Ding Hou	CD	Rhizophoraceae	57.5
Ceriops tagal (Pierre.) Robins	CT	Rhizophoraceae	2.5
Excoecaria agallocha L	EA	Euphorbiaceae	42.5
Heritiera fomes Buch. Ham	HF	Sterculiaceae	10.0
Lumnitzera racemosa Willd	LR	Combretaceae	2.5
Xylocarpus mekongensis Pierre	XM	Meliaceae	5.0
Shrubs			
Aegialitis rotundifolia Roxb	AR	Plumbaginaceae	52.5
[b]*Phoenix paludosa* Roxb	PP	Arecaceae	27.5
[a]*Dalbergia spinosa* Roxb	DS	Papilionaceae	10.0
Herbs			
[a]*Acanthus ilicifolius* L	AI	Acanthaceae	57.5
[a]*Suaeda maritima* Dumort	SM	Chenopodiaceae	12.5
[a]*Sesuvium portulacastrum* L	SP	Aizoaceae	17.5
Climbers			
[a]*Caesalpinia bonduc* (L.) Roxb	CB	Caesalpiniaceae	7.5
[a]*Caesalpinia crista* L	CC	Caesalpiniaceae	5.0
[a]*Derris trifoliata* Lour	DT	Papilionaceae	32.5

[a]Mangrove associates/back mangrove (Naskar 2004)
[b]Monocot species

3.2 *Effects of Tidal Inundation*

Among the species recorded, *Avicennia alba* and *Acanthus* preferred diurnal inundation while *Brownlowia*, *Bruguiera*, *Heritiera fomes*, *Lumnitzera* and *Xylocarpus granatum* were restricted to spring tides. However, tidal inundation significantly affected the density of five species, i.e. *Avicennia marina* ($F_{0.05(3,36)} = 3.709$), *Ceriops decandra* ($F_{0.01(3,36)} = 13.507$), *Excoecaria* ($F_{0.05(3,36)} = 3.685$), *Phoenix paludosa* ($F_{0.05(3,36)} = 3.638$) and *Suaeda maritime* ($F_{0.01(3,36)} = 12.809$). *Avicennia marina* exhibited low density under diurnal inundation while very high density with stunted growth under no inundation. *Suaeda* was also highly dense under no inundation while absent under diurnal inundation. *Phoenix* was restricted to and *Excoecaria* preferred spring tides. *Ceriops decandra* exhibited very high density under summer spring tides and spring tides compared to that under diurnal or no tides.

Soil parameters affecting vegetation were influenced by the frequency and duration of tidal inundation. Clay, organic carbon, nitrogen and sodium displayed higher values under diurnal inundation while pH, sand, potassium, calcium and magnesium at summer spring tides or no tides. Only two parameters, soil magnesium ($F_{0.05(3,36)}$ = 3.560) and phosphorus ($F_{0.05(3,36)}$ = 3.452), varied significantly under different tidal categories. Soil phosphorus was significantly higher under diurnal inundation while soil magnesium under summer spring tides.

3.3 *Effects of Edaphic and Biotic Parameters*

The single-linkage hierarchical classification of species densities generated seven clusters (Fig. 2). *Aegialitis, Avicennia alba, Brownlowia* and *Caesalpinia crista* separated as cluster 1 to 4 of single species each, respectively, hence, unaffected by the presence or absence of other species. Cluster 5 consisted of three species—*Heritiera, Lumnitzera* and *Phoenix*. Cluster 6 comprised six species—*Acanthus, Avicennia marina, Bruguiera, Dalbergia spinosa, Derris trifoliata* and *Sesuvium portulacastrum*. Cluster 7 got the maximum number of species—*Aegiceras, Avicennia officinalis, Ceriops decandra, C. tagal, Caesalpinia bonduc, Excoecaria, Suaeda* and *Xylocarpus*.

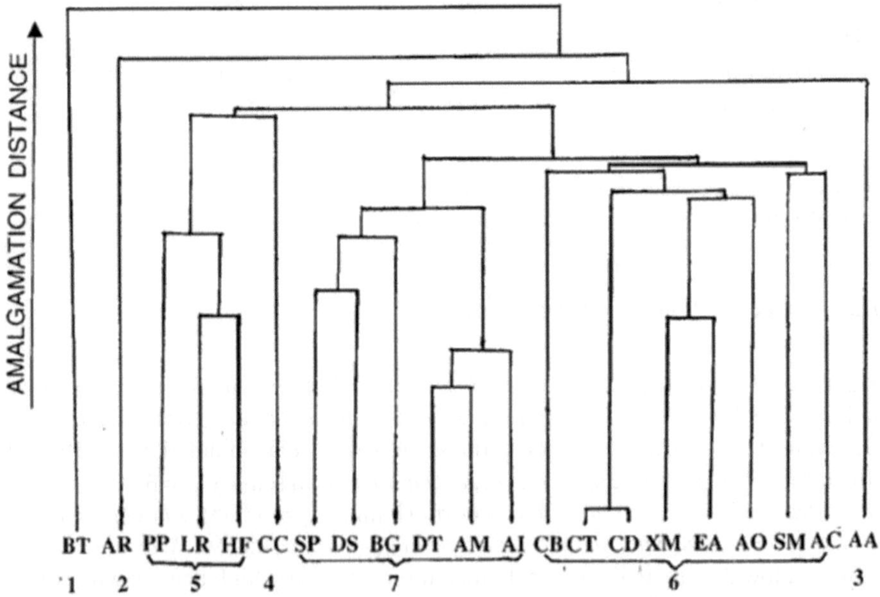

Fig. 2 Dendrogram showing seven species clusters. See Table 1 for species names

Most of the correlations between species density and soil parameters were non-significant, except a few like *Avicennia alba* with calcium, *A. marina* with phosphorus and salinity, *Aegiceras* with nitrogen, and *Ceriops decandra* with nitrogen and organic carbon. However, most of the soil variables were significantly correlated with each other like pH was positively correlated with sand and magnesium but negatively with organic carbon and nitrogen, and salinity positively with silt, clay, sodium, potassium, phosphorus, calcium and magnesium. Table 2 displays regression equations predicting species densities. The independent variables significantly explained total variation of 42% or above in the species densities, except for *Avicennia officnalis* and *Suaeda* (non-significant). In clusters 1 and 2, soil calcium explained the maximum variation in *Avicennia alba* (14.5% out of total 49.8%) and sodium in *Aegialitis* (32.5% out of total 46.2%). Among physical parameters, salinity explained the maximum variation in *Avicennia alba* (8%) and *Aegialitis* (4.6%) out of total variation.

In clusters 5, 6 and 7, biotic parameters accounted for a huge fraction of total variation. In cluster 5, the occurrence of *Heritiera* was highly explained by *Lumnitzera* (41.88%) and *Phoenix* (15.20%) out of total 64.6% variation. *Heritiera* explained huge variation in *Phoenix* (25%) out of total 45.4% variation. Among soil nutrients, sodium explained maximum variation in *Heritiera* (1.4% out of total 64.6%), interacting negatively. Phosphorus highly explained the occurrence of *Phoenix* (4.2%) from total variation of 45.4%. Among soil physical parameters, sand had negative influence on *Heritiera* (2.9% variation) and *Phoenix* (3.3% variation).

In cluster 6, the occurrence of *Avicennia marina* was highly explained by climber *Derris* (58.06%) and herb *Sesuvium* (10.14%) out of total 77.22% variation explained. *Derris* and *Dalbergia* accounted for maximum variation in *Acanthus*, explaining 49.23% and 16.50% out of total 78.08% variation, respectively. The occurrence of *Dalbergia* was highly explained by *Sesuvium* (36.39%) and *Bruguiera* (29.63%) out of total 82.47% variation. Out of total 66% variation in *Sesuvium*, *Dalbergia* explained 36.39% and *Bruguiera* 17.63%. Variation in *Derris* was highly explained by *Avicennia marina* (58.06%), *Acanthus* (8%) and *Bruguiera* (1.54%) out of total 82.4% variation.

Among soil nutrients, calcium explained 1.5% variation in *Avicennia marina* and 1% variation in *Acanthus* out of total 77.2% and 78.1% variation, respectively. Phosphorus explained 4.1% variation in *Sesuvium* and 1.1% variation in *Dalbergia* out of 66% and 82.5% variation, respectively. Organic carbon influenced *Derris* positively (3.3% out of total 82.4% variation). Among physical parameters, salinity explained 1% variation in *Avicennia marina* and 0.3% in *Derris* out of the total variations. pH explained maximum variation in *Dalbergia* (2% out of 82.5%). Sand and clay explained 7.8% and 2.5% variation from total variations in *Acanthus* and *Sesuvium*, respectively.

In cluster 7, *Ceriops decandra* positively interacted with *Aegiceras*, *Avicennia officinalis*, *C. tagal* and *Excoecaria* but *C. tagal* alone explained 96.43% variation out of total 97.07% variation in *Ceriops decandra.* Out of total 84.5% variation in *Excoecaria*, *Xylocarpus* explained 42.7% and *Ceriops decandra* 21.8%. The occurrence of *Avicennia officinalis* was highly explained by *Xylocarpus* (20%) and *Excoecaria* (3.9%) out of total 43% variation. *Suaeda* and *Caesalpinia bonduc* highly

Table 2 Multiple regression equations predicting species densities on the basis of edaphic and biotic parameters. The linear regression equations were of the order: $Y = a_1 + b_1X_1 + b_2X_2 + \cdots b_nX_n \pm SE$; where Y = dependent variable (species density), a = Y intercept, b = partial regression coefficient, X = independent variable, SE = standard error of estimate, R^2 = coefficient of determination (For species names see Table 1)

Y	a_1	$b_1x_1 + b_2x_2 + \cdots b_nx_n$	SE	R^2
AA	−11067.35	41.67(Ca) +21.32(sand) +67.37(N)+ 17.55(Mg) −1.55(Na) +64.77(sal)+ 930.43(pH) +711.99(OC) +2.67(K)	847.80	0.4982
AM	−7647.98	0.36(DT) −0.00(SP)+12.58(DS) +33.7(Mg) −8.10(BG)+44.96(silt) −103.08(sal)+ 43.05(Ca) +5.62(K)+734.78(pH)	1608.93	0.7722
AO	−162442.53	200.88(XM)+24049.94(pH) −427.32(Mg) −2.40(EA) −14.01(Na) −501.74(P)+ 58.05(K)+13667.99(OC) −0.03(SM)+ 288.09(sal)+0.22(CB)+19.12(CD)+ 0.15(AC) −2.40(CT)	14583.50	0.4298
AC	93296.70	173.81(P)+0.01(SM) −0.20(CB) −21229.59(OC)+8.98(Na) −11698.61(pH) −0.84(EA) −266.88(clay) −97.77(Ca) −2.41(CT)+162.38(N)+20.25(XM)+ 12.91(CD)+111.25(silt)+62.89(Mg) −10.48(K)	4846.96	0.6555
CD	−302.17	0.17(CT)+0.09(Ca)+103.02(OC) −1.73(N) −0.05(Na)+1.71(clay)+0.01(EA) −0.0009(CB)+0.001(AC)+ 41.26(pH)-0.17(XM)	50.13	0.9707
EA	21690.05	37.39(XM)+4.81(CD)+0.05(CB) −76.48(Ca) −62.48(P) −0.03(AO) −40.23(Mg) −5.90(N) −0.01(SM) −0.08(AC) −2823.64(OC)+57.64(sal) −1532.63(pH) −74.21(clay) −40.75(sand)	1594.78	0.8450
HF	650.81	3.31(LR)+0.01(PP) −57.09(pH)-0.08(Na) −1.95(sand)+0.36(Mg)+ 69.17(OC)-2.10(N)+1.81(P) −0.18(K)	100.68	0.6457
AR	20598.59	38.80(Na) −1049.32(sal)+693.56(Ca) −9470.11(pH)+229.63(sand) −312.16(P)+ 5406.99(OC)	17635.10	0.4620
DS	402.64	0.0002(SP)+0.62(BG)+0.13(AI)-50.85(pH) −1.61(P) −0.01(DT)+0.01(AM)+2.18(sal) −0.16(K)	43.41	0.8247
PP	56321.88	17.06(HF) −125.74(P) −51.93(LR) −249.78(clay) −174.27(sand) −77.48(Mg) −146.33(N) −3534.39(pH) −16.57(K)	4631.27	0.4536
AI	−406.50	0.03(DT)+1.64(DS)+1.82(sand) −0.65(BG) −4.59(Ca) −3.36(clay) −2.69(Mg)+ 92.53(pH)+0.04(Na) −1.59(N) −0.01(AM)	152.11	0.7808

(continued)

Table 2 (continued)

Y	a_1	$b_1x_1 + b_2x_2 + \cdots b_nx_n$	SE	R^2
SM	45162.54	1.91(AC) + 1650.85(silt) −1.28(CT) + 5394.64(sal) −2648.82(Ca) −39.83(Na) −1099.14(Mg) −63556.47(OC) −1516.44(P) +532.76(N) + 419.73(XM) −7.79(EA) −0.56(AO) + 67.42(CD)	57155.51	0.4040
SP	−113466.69	2132.44(DS) −1346.35(BG) + 4296.05(P) + 2781.62(Mg) −4.80(DT) −150644.05(OC) +5274.78(clay) −4993.84(sal)	158953.78	0.6603
DT	−3727	1.14(AM) + 10.33(AI) −26.88(DS) + 3861.1(OC) + 10.47(BG) + 62.17(clay) −0.0004(SP) −126.35(N) −63.62(Mg) +754.95(pH) −10.33(K) + 109.44(sal) −45.41(Ca)	2901.96	0.8242

explained the occurrence of *Aegiceras* (13.05% and 6.5%, respectively, out of total 65.6% variation), while the occurrence of *Suaeda* was highly explained by *Aegiceras* (16.64% out of 40.4%).

Among soil nutrients, calcium interacted negatively with *Excoecaria* and *Suaeda*, explaining 4.5% variation in each out of total 84.5% and 40.4% variations, respectively. Phosphorus explained the occurrence of *Aegiceras* (18.2%) and *Excoecaria* (1.8%) from total variations of 65.6% and 84.5%, respectively. The occurrence of *Avicennia officinalis* was highly explained by magnesium (4% variation out of total 43%). Calcium and nitrogen each explained 0.1% variation in *Ceriops decandra* from total 97.1% variation. Among physical parameters, pH explained 3.8% and 4.28% variation from total variation of 65.6% and 43%, in *Aegiceras* and *Avicennia officinalis*, respectively. Salinity explained 0.4% variation in *Excoecaria* from total 84.5%. Silt explained 3.4% variation out of 40.4% in *Suaeda*. Clay explained 0.1% variation out of 97.1% in *Ceriops decandra*.

3.4 Species Distribution Maps

Figures 3 and 4 show the distribution of soil parameters in the Lothian Island. Soil salinity was high in northern and middle portions, and extremely high (28–35 ppt) in the northern tip. The southern-ridged portion had lower values. Important soil nutrients like nitrogen and phosphorus and cations like sodium, calcium and magnesium also concentrated in the northern half of the island. Maps predicting density distribution of some species are shown in Figs. 5 and 6. *Avicennia alba* was dense in the northern tip. *A. marina* occurred all over the island but very dense near the north-eastern coast and low in the middle creek-intersected portion. *A. officinalis* occurred all over the island. *Aegiceras* was abundant in the northern and middle portion of the island. *Ceriops decandra* and *Excoecaria* were concentrated in the southern portion.

Shrubs *Aegialitis* and *Phoenix* displayed high density in the middle portion. The herb *Acanthus* was very dense in the northern tip. *Sesuvium* and *Suaeda* were abundant in the northeastern portion followed by middle and southern portion. Climber *Derris* was mainly concentrated in the northeastern and southern coastal regions.

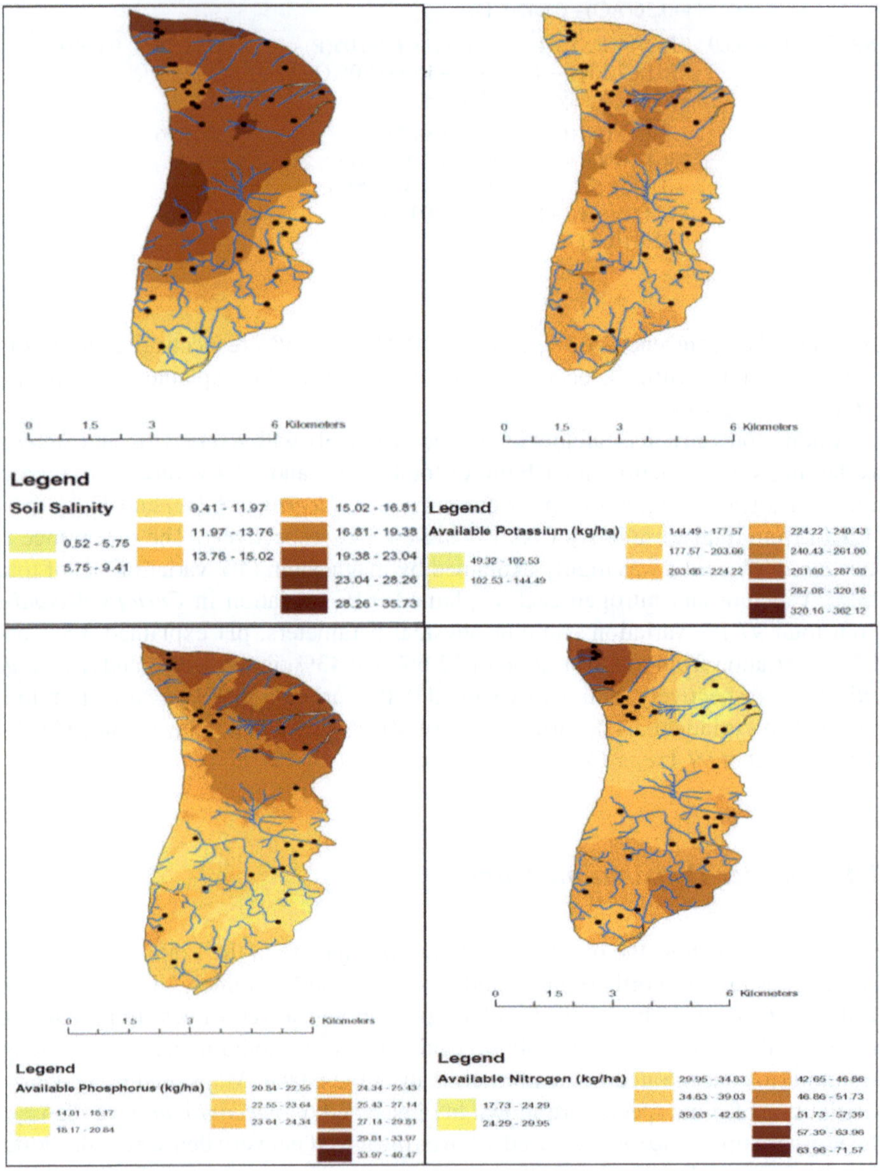

Fig. 3 Prediction maps showing the distribution of soil salinity (ppt) and NPK concentration (kg/ha) over Lothian Island

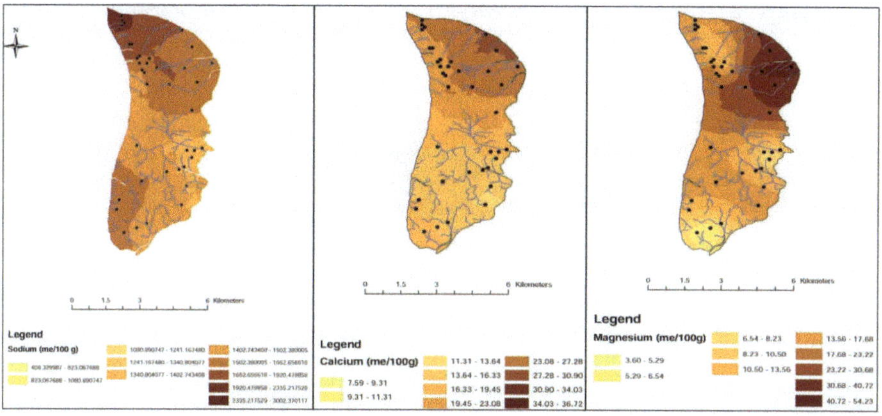

Fig. 4 Prediction maps showing distribution of some soil cations (me/100 g) over Lothian Island

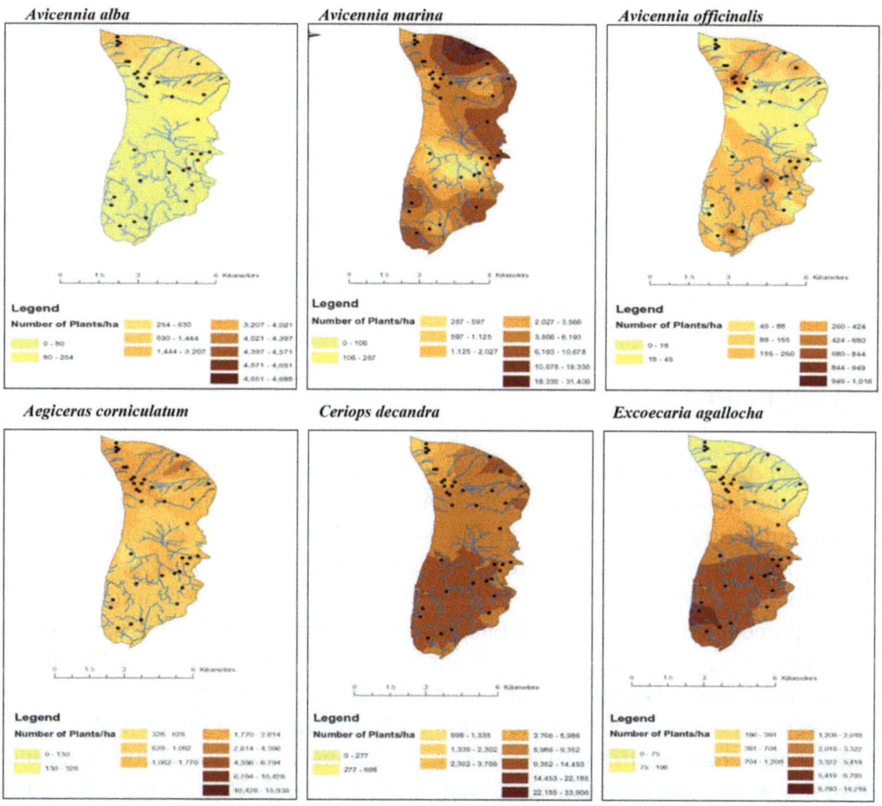

Fig. 5 Prediction maps showing distribution of tree species (N/ha) species over Lothian Island

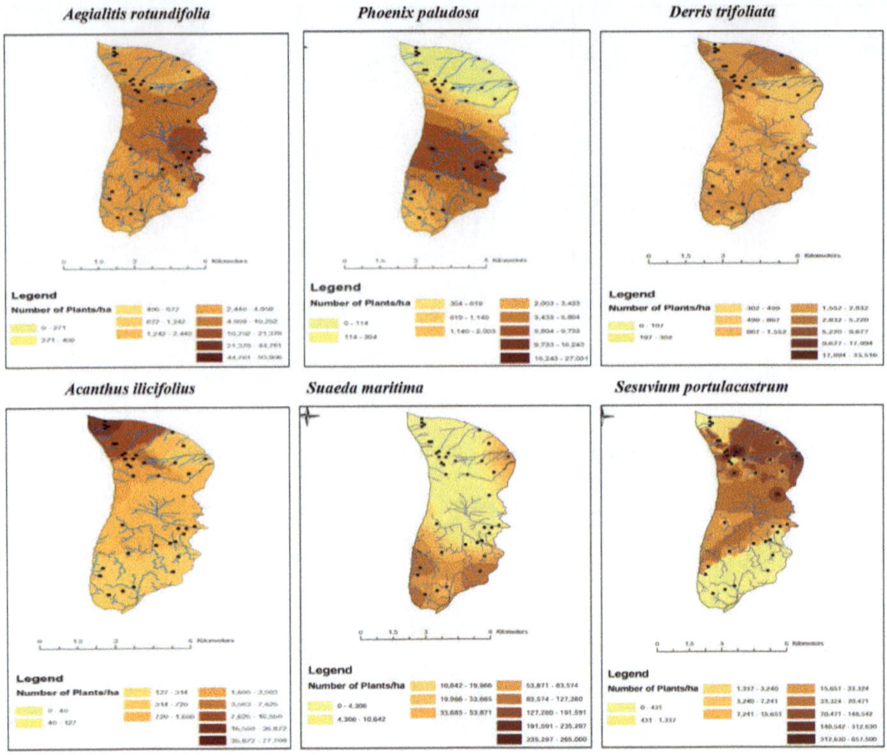

Fig. 6 Prediction maps showing distribution of shrubs, climber and herbs over Lothian Island

4 Discussion

The present investigation revealed new findings regarding mangrove species distribution in thc Sundarbans. *Avicennia alba*, *A. marina* and *Acanthus* dominated under diurnal inundation while *Excoecaria*, *Phoenix*, *Ceriops* and *Derris* dominated under spring tides. This is similar to earlier reports from this region (Pal et al. 1996; Saha and Choudhury 1995). *Heritiera, Bruguiera* and *Xylocarpus* were infrequent and restricted under spring tides. Tidal factors have been responsible for disappearance of *Heritiera fomes* in the Indian Sundarbans (Balsco 1977; Mukherjee and Mukherjee 1978). The zonation and succession of mangroves are reported to correlate with the frequency and duration of tidal inundation, although modified by local topography (Ball 1996; Semeniuk 1983). *Avicennia* was the pioneer species to colonize sandy substrates at the Beechwood area, Durban Bay (Naidoo 1980). In Australian mangroves, association of *Avicennia* and *Aegiceras corniculatum* had strong downstream water's edge affinities, while association of *Excoecaria agallocha* and several others had opposite tendencies, i.e. closest to the land (Bunt and Williams 1981).

Tidal inundation, influencing soil saturation hence redox potential, has major implications for both micro- and macro-nutrients (Boto 1982; Valiela 1984). Many studies reported that tidal flooding affects soil–vegetation relationship through different soil nutrients—chloride concentrations and organic carbon (Ukpong 1998; Boto et al. 1989), availability of nitrogen (Boto and Wellington 1983, 1984; Middelburg et al. 1996) and phosphorus (Valiela 1984). In this study, soil phosphorus was more available under diurnal inundation. It might be due to the presence of abundant organic matter in anaerobic soils may contribute to the accumulation of dissolved phosphate by inhibiting its adsorption onto clay surface (Valiela 1984). Under aerobic conditions, phosphorus readily adsorbs on clay and precipitates with calcium, magnesium and iron becoming largely unavailable.

In the present investigation, three species clusters, each comprising of a few species, demonstrated significant positive or negative interaction among themselves. Phytotoxic exudates, shading and associated growth, tidal sorting of propagule and physiological tolerance are some ways of species interactions (Clarke and Hannon 1971; Rabinowitz 1978). Correlations among the soil variables suggested that complex interactions among soil variables determine the species composition. For example, soil pH was associated with a number of soil parameters. Soil pH controls the chemical status and mobility of many elements and significantly affects plant growth due to change in availability of both essential elements such as phosphorus (P) and non-essential elements such as aluminium (Al) that can be toxic to plants at elevated concentrations (Black 1993; Slattery et al. 1999). The primary nutrients (N, P, K) as well as the secondary nutrients (S, Ca, Mg) are more available at pH 5.5 for organic and 6.5 for mineral soils, than at any other pH (Donahue et al. 1985). With increase in soil pH, we observed decrease in soil nitrogen while phosphorus increased though insignificantly. In flooded anaerobic soils with high pH, nutrient availability can be decreased markedly by denitrification or volatilization losses of inorganic nitrogen, alternatively phosphate mobility may be greater (De Laune et al. 1976; Sample et al. 1980).

Salinity displayed positive correlation with exchangeable bases like sodium, potassium, calcium and magnesium. Concentration of sodium was also higher than potassium in Lothian soil—reverse of that reported for less saline eastern islands of Indian Sundarbans with higher values of soil potassium than sodium (Nandy Datta 2003). Excess Na^+ might cause nutrient deficiency by reducing the amounts of available K^+, Mg^+ and Ca^+ (Epstein 1972), and interferes with the function of K^+ as a cofactor in various reactions (Hseu and Chen 2000). *Rhizophora apiculata* formed the seaward fringe in eastern islands of the Indian Sundarbans (Nandy Datta 2003), and displayed better growth in plantation sites enriched with potassium (Kathiresan et al. 1994a). This association of *Rhizophora* with potassium might be a probable reason for its complete absence in Lothian Island. Difference in salinity in the eastern and western islands of Indian Sundarbans is reported to control distribution of at least some mangrove species—*Avicennia officinalis*, *A. alba* and *Acanthus ilicifolius* dominate in high salinity while *Bruguiera gymnorrhiza*, *Aegiceras majus*, *Excoecaria agallocha*, *Heritiera minor* and *Phoenix paludosa* in low salinity (Matilal et al. 1986; Pal et al. 1996). We also had similar findings. The leaves of *Avicen-*

nia alba, *A. marina* and *Acanthus ilicifoilius*, from highly saline soils of Lothian and Prentice islands, Sundarbans, were rich in sodium content, whereas those of *Excoecaria agallocha* and *Phoenix paludosa* from Sagar and Harinbari islands were low (Sah et al. 1989). In the Bangladesh Sundarbans also, regional occurrence of *Ceriops decandra*, *Excoecaria agallocha*, *Heritiera fomes* and *Sonneratia apetala* was affected by regional variation in salinity (5–15 ppt) (Ellison et al. 2000).

5 Conclusions

The present study provides useful information regarding the distribution of mangrove species in the Indian Sundarbans. Mangrove species responded differently to different tidal regimes. Regression analysis accounted for 42% or more variation in species densities. *Heritiera fomes* and *Avicennia alba* had narrow distribution. More selectiveness in site location for reforestation is therefore required for these species. *Bruguiera gymnorrhiza*, *Ceriops tagal*, *Lumnitzera racemosa* and *Xylocarpus granatum* were recorded from single site each, and hence no significant conclusion could be drawn regarding their preferences for soil variables. *Avicennia marina* seemed insensitive to the variations of the soil properties and occurred all over the island. *Avicennia officinalis* and *Ceriops* also displayed wide amplitudes. Prediction maps displayed a northern seaward fringe of *Avicennia alba* and herb *Acanthus* under high soil salinity. *Ceriops*, *Excoecaria*, *Aegialitis* and *Phoenix* concentrated in the middle and southern portion of the island with lower soil salinity.

Acknowledgements We thank Professors A. Choudhuri, D. Roy and S. Bandopadhyay of Indian Statistical Institute, Kolkata for statistical assistance. We are also thankful to the Conservator and Joint Director, Sundarbans Biosphere Reserve, and DFO, South 24 Parganas, West Bengal for the necessary permission and help they provided for the fieldwork. Last, but not the least, we thank the field assistants for their sincere help.

References

Alongi, D. M., Christofferson, P., & Tirendi, F. (1993). The influence of forest type on microbial-nutrient relationships in tropical mangrove sediments. *Journal of Experimental Marine Biology and Ecology, 171,* 201–223.

Ball, M. C. (1996). Comparative ecophysiology of mangrove forest and tropical lowland moist forest. In S. S. Mulkey, R. L. Chazdon, & A. O. Smith (Eds.), *Tropical forest plant ecophysiology* (pp. 461–469). New York: Chapman and Hall.

Ball, M. C. (1998). Mangrove species richness in relation to salinity and waterlogging: A case study along the Adelaide River floodplain, northern Australia. *Global Ecology & Biogeography Letters, 7,* 73–82.

Banerjee, L. K. (1987). Ecological studies on the mangals in the Mahanadi estuarine-delta Orissa: India. *Tropical Ecology, 28,* 117–125.

Banerjee, L. K, Sastry, A. R. K., Nayar, M. P. (1989). Mangroves in India, identification manual. *Botanical survey of India, Kolkata*.

Baruah, T. C., & Barthakur, H. P. (1997). *A textbook of soil analysis*. New Delhi: Vikas Publishing House Pvt. Ltd.
Black, C. A. (1993). *Soil fertility evaluation and control*. Boca Raton, FL.: Lewish Publishers.
Blasco, F. (1977). Outlines of ecology, botany and forestry of the mangals of Indian Subcontinent. In V. J. Chapman (Ed.), *Wet coastal ecosystems. Ecosystems of the world. 1* (pp. 241–257). Elsevier Scientific Publishing Company, Amsterdam.
Boto, K. G. (1982). Nutrient and organic fluxes in mangroves. In B. F. Clough (Ed.), *Mangrove ecosystems in Australia* (pp. 239–257). Canberra: Australian National University Press.
Boto, K. G., Alongi, D. M., & Nott, A. L. (1989). Dissolved organic carbon-bacteria interactions at sediment-water interface in a tropical mangrove ecosystem. *Marine Ecology Progress Series, 51,* 243–251.
Boto, K. G., & Wellington, J. T. (1983). Phosphorus and nitrogen nutritional status of a northern Australian mangrove forest. *Marine Ecology Progress Series, 11,* 63–69.
Boto, K. G., & Wellington, J. T. (1984). Soil characteristics and nutrient status in a northern Australian mangrove forest. *Estuaries, 7,* 61–69.
Bunt, J. S., & Williams, W. T. (1981). Vegetational relationships in the mangroves of tropical Australia. *Marine Ecology Progress Series, 4,* 349–359.
Chapman, V. J. (1976). *Mangrove vegetation*. Vaduz, Germany: J. Cramer.
Chen, R., & Twilley, R. R. (1998). A gap dynamic model of mangrove forest development along gradients of soil salinity and nutrient resources. *Journal of Ecology, 86,* 37–52.
Chen, R., & Twilley, R. R. (1999). Patterns of mangrove forest structure and soil nutrient dynamics along the Shark River estuary, Florida. *Estuaries, 22,* 955–970.
Cintron, G., & Schaeffer-Novelli, Y. (1984). Methods of studying mangrove structure. In S. C. Snedaker & J. C. Snedaker (Eds.), *The Mangrove ecosystem: Research methods* (pp. 91–113). Paris: UNESCO.
Clarke, L. D., & Hannon, N. (1971). The mangrove swamp and salt marsh communities of the Sydney district IV, significance of species interaction. *Journal of Ecology, 59,* 535–553.
Dahdouh-Guebas, F., Di Nitto, J. L., Bosire, D., Lo Seen, J. O., & Koedam, D. N. (2005). How effective were mangroves as a defense against the recent tsunami. *Current Biology, 15,* 443–447.
De Laune, R. D., Patrick Jr, W. H., & Brannon, J. M. (1976). Nutrient transformations in Louisiana salt marsh soils. Baton Rouge, La., Center for Wetland Resources, Louisiana State University (Sea Grant publication, LSU-T-76–009).
Donahue, R. L., Miller, R. W., Schickluna, S. C. (1985). *An introduction to soils and plant growth*. Engelwood Cliffs, New Jersey: Prentice-Hall Inc.
Ellision, A. M., Mukherjee, B. B., & Karim, A. (2000). Testing patterns of zonation in mangroves: Scale dependence and environmental correlates in the Sundarbans of Bangladesh. *Journal of Ecology, 8,* 813–824.
Epstein, E. (1972). *Mineral nutrition of plants: principles and perspectives*. New York: Wiley.
Feller, I. C. (1995). Effects of nutrient enrichment on growth and herbivory of dwarf red mangrove *Rhizophora mangle*). *Ecological Monographs, 54,* 477–505.
Feller, I. C., McKee, K. L., Whigham, D. F., & O'Neill, J. P. (2003). Nitrogen vs. phosphorus limitation across an ecotonal gradient in a mangrove forest. *Biogeochemistry, 62,* 145–175.
Forest Survey of India. (1999). The state of forest report. Forest Survey of India, Ministry of Environment and Forests, Dehradun.
Giri, C., Pengra, B., Zhu, Z., Singh, A., & Tieszen, L. L. (2011). Monitoring mangrove forest dynamics of the Sundarbans in Bangladesh and India using multi-temporal satellite data from 1973–2000. *Estuarine Coastal Shelf Science, 73*(1–2), 91–100.
Hosakawa, T., Tagawa, H., & Chapman, V. J. (1977). Mangals of Micronesia, Taiwan, Japan, the Philippines and Oceania. In V. J. Chapman (Ed.), *Ecosystems of the world. I wet coastal ecosystems* (pp. 271–290). Elsevier Scientific, Amsterdam.
Hussain, S. A., & Badola, R. (2010). Valuing mangrove benefits: Contribution of mangrove forests to local livelihoods in Bhitarkanika Conservation Area, East Coast of India. *Wetlands Ecology and Management, 18,* 321–331.

Hseu, Z. Y., & Chen, Z. S. (2000). Monitoring the changes of redox potential, pH and electrical conductivity of the mangrove soils in northern Taiwan. *Proceedings of National Science Council ROC (B), 24,* 143–150.
Jackson, M. L. (1973). *Soil chemical analysis*. New Delhi: Prentice Hall of India Pvt. Ltd.
Joshi, H., & Ghose, M. (2002). Structural variability and biomass production of mangroves in Lothian island of Sundarbans, India. In S. Javed & A. G. de Soyza (Eds.), *Research and management options for Mangrove and Saltmarsh ecosystems* (pp. 146–158). Abu Dhabi, UAE: ERWDA.
Joshi, H., & Ghose, M. (2003). Forest structure and species distribution along soil salinity and pH gradient in mangrove swamps of the Sundarbans. *Tropical Ecology, 44,* 195–204.
Gupta Joshi, H., & Ghose, M. (2014). Community structure, species diversity and aboveground biomass of the Sundarbans mangrove swamps. *Tropical Ecology, 55*(3), 283–303.
Kathiresan, K., Moorthy, P., & Rajendran, N. (1994). Seedling performance of mangrove *Rhizophora apiculata* (Rhizophorales: Rhizophoraceae) in different environments. *Indian Journal of Marine Sciences, 23,* 168–169.
Krauss, K. W., Lovelock, C. E., McKee, K. L., Lo'pez-Hoffmen, L., Ewe, S. M. L., & Sousa, W. P. (2008). Environmental drivers in mangrove establishment and early development: A review. *Aquatic Botany, 89,* 105–127.
Lee, S. Y. (1999). Tropical mangrove ecology: Physical and biotic factors influencing ecosystem structure and function. *Australian Journal of Ecology, 24,* 355–366.
Lugo, A. E. (1980). Mangrove ecosystems: Successional or steady state? Tropical succession. *Biotropica Supple, 12,* 65–72.
Lugo, A. E., & Snedaker, S. C. (1974). The ecology of mangroves. *Annual Review of Ecology and Systematics, 5,* 39–64.
MacGuiness, K. A. (1997). Seed predation in a tropical mangrove forest: A test for the dominance predation model in northern Australia. *Journal of Tropical Ecology, 13,* 293–302.
Mall, L. P., Singh, V. P., Garge, A., & Pathak, S. M. (1987). Ecological studies on mangrove forests of Ritchie's Archipelago in relation to substrata. *Tropical Ecology, 28,* 182–197.
Matilal, S., Mukherjee, B. B., Chatterjee, N., & Gupta, M. D. (1986). Studies on soil & vegetation of mangrove forests of Sundarbans. *Indian Journal of Marine Sciences, 15,* 181–184.
McKee, K. L. (1993). Soil physicochemical properties and mangrove species distribution—Reciprocal effects? *Journal of Ecology, 81,* 477–487.
Middelberg, J. J., Nieuwenhuize, J., Slim, F. J., & Ohowa, B. (1996). Sediment biogeochemistry in an East African mangrove forest (Gazi Bay, Kenya). *Biogeochemistry, 34,* 133–155.
Mukherjee, B. B., & Mukherjee, J. (1978). Mangroves of Sundarbans, India. *Phtyomorphology, 28,* 177–192.
Naidoo, G. (1980). Mangrove soils of the Beachwood area, Durban. *Journal of South African botany, 46,* 293–304.
Nandy Datta, P., & Ghose, M. (2003). Estimation of osmotic potential and free amino acids in some mangroves of the Sundarbans, India. *Acta Botanica Croatica, 62,* 37–45.
Naskar, K. R., & Guha-Bakshi, D. N. (1987). *Mangrove swamps of the Sundarbans: An ecological perspective*. Kolkata, India: Naya Prokash.
Naskar, K. R. (2004). *Manual of Indian mangroves* (p. 220). Delhi: Daya Publishing House.
Nazim, K., Ahmed, M., Shaukat, S. S., Khan, M. U., & Ali, Q. M. (2013). Age and growth rate estimation of grey mangrove Avicennia Marina (Forsk.) Vierh from Pakistan. *Pakistan Journal of Botany, 45*(2), 535–542.
Nguyen, H. T., Stanton, D. E., Schmitz, N., Farquhar, G. D., & Ball, M. C. (2015). Growth responses of the mangrove Avicennia marina to salinity: Development and function of shoot hydraulic systems require saline conditions. *Annals of Botany, 115,* 397–407.
Nickerson, N. H., & Thibodeau, F. R. (1985). Association between pore water sulfide concentrations and the distribution of mangroves. *Biogeochemistry, 1,* 183–192.
Okimoto, Y., Nose, A., Ikeda, K., Agarie, S., Oshima, K., Tateda, Y., et al. (2008). An estimation of CO2 fixation capacity in mangrove forest using two methods of CO_2 gas exchange and growth curve analysis. *Wetlands Ecology and Management, 16,* 155–171.

Olsen, S. R., Cole, C. V., Watanabe, F. S., & Dean, L. A. (1954). Estimation of available phosphorus in soils by extraction with sodium bicarbonate. U.S. Department of Agriculture Circular 939.

Pal, D., Das, A. K., Gupta, S. K., & Sahoo, A. K. (1996). Vegetation pattern and soil characteristics of some mangrove forest zones of the Sundarbans, West Bengal. *Indian Agriculturist, 40,* 71–78.

Piper, C. S. (1960). *Soil and plant analysis*. New York: Inter Science Publishers.

Rabinowitz, D. (1978). Early growth of mangrove seedlings in Panama, and a hypothesis concerning the relation of dispersal and zonation. *Journal of Biogeography, 5,* 113–133.

Reef, R., Feller, I. C., & Lovelock, C. E. (2010). Nutrition of mangroves. *Tree Physiology, 30,* 1148–1160.

Sah, K. D., Sahoo, A. K., Gupta, S. K., & Banerjee, S. K. (1989). Mangrove vegetation of Sundarbans and their effect on the physico-chemical and nutrient status of the soils. *Proceedings of the Indian National Science Academy B, 55,* 125–132.

Saha, S., & Choudhury, A. (1995). Vegetation analysis of restored and natural mangrove forests in Sagar Island, Sundarbans, East Coast of India. *Indian Journal of Marine Sciences, 24,* 133–136.

Sample, E. C., et al. (1980). Reactions of phosphate fertilizers in soils. In F. E. Khasawnch, et al. (Eds.), *The Role of phosphorus in agriculture*. Madison, WI: American Society of Agronomy, Crop Science Society America, Soil Science Society America.

Semeniuk, V. (1983). Mangrove distribution in northwestern Australia in relationship to regional and local freshwater seepage. *Vegetatio, 53,* 11–31.

Slattery, W. J., Conyers, M. K., & Aitken, R. L. (1999). Soil pH, aluminium, manganese, and lime requirement. In K. I. Peverill, L. A. Sparrow, & D. J. Reuter (Eds.), *Soil analysis: An interpretation manual* (pp. 103–128). Collingwood, Australia: CSIRO Publishing.

Smith, T. J., III. (1987). Seed predation in relation to tree dominance and distribution in mangrove forests. *Ecology, 68,* 266–273.

Smith, T. J., III. (1988). Differential distribution between subspecies of the mangrove Ceriops tagal: Competitive interactions along a salinity gradient. *Aquatic Botany, 32,* 79–89.

Smith, T. J., III. (1992). Forest Structure. In A. I. Robertson & D. M. Alongi (Eds.), *Tropical Mangrove Ecosystems* (pp. 101–136). New York: American Geophysical Union.

Subiah, B. V., & Asiza, G. L. (1956). A rapid procedure for the estimation of available nitrogen in soils. *Current Science, 25,* 259–261.

Tomlinson, P. B. (1986). *Botany of Mangroves*. Cambridge: Cambridge University Press.

Ukpong, I. E. (1991). The performance and distribution of species along soil salinity gradients of mangrove swamps in southeastern Nigeria. *Vegetatio, 95,* 63–70.

Ukpong, I. E. (1998). The composition and distribution of species in relation to soil nutrient gradients in mangrove swamps in South Eastern Nigeria. *Tropical Ecology, 39,* 55–67.

Valiela, I. (1984). *Marine ecological processes*. New York: Springer.

Vo, Q. T., Kuenzer, C., Vo, Q. M., Moder, F., & Oppelt, N. (2012). Review of valuation methods for mangrove ecosystem services. *Ecological Indicators, 23,* 431–446.

Wakushima, S., Kuraishi, S., & Sakurai, N. (1994a). Soil salinity and pH in Japanese mangrove forests and growth of cultivated mangrove plants in different soil conditions. *Journal of Plant Research, 107,* 39–46.

Wakushima, S., Kuraishi, S., Sakurai, N., Supappibul, K., & Siripatanadilok, S. (1994b). Stable soil pH of Thai mangroves in dry and rainy seasons and its relation to zonal distribution of mangroves. *Journal of Plant Research, 107,* 47–52.

Walkely, A., & Black, I. A. (1934). An examination of the Degtjareff method for determining soil organic matter and a proposed modification of the chromic acid titration method. *Soil Science, 34,* 54–56.

Walsh, G. E. (1974). Mangroves: A review. In R. Reimold & W. Queen (Eds.), *Ecology of halophytes* (pp. 51–174). New York: Academic Press.

Walter, H. (1977). Climate. In V. J. Chapman (Ed.), *Ecosystems of the world* (pp. 61–67). Elsevier, New York: Wet coastal ecosystems.

Watson, J. (1928). Mangrove forests of the Malay Peninsula. *Malayan Forest Records 6*. Singapore: Fraser & Neave, Ltd..

Characterization of Extended Uniform Distribution and Its Applications

Ratan Dasgupta

Abstract Extended uniform distributions $G(y) = (y/\theta)^\alpha, \alpha > 0, \theta > 0, y \in (0, \theta]$, and its discrete version have applications in modeling random variables related to growth data, discrete and continuous (Dasgupta 2017). Notion of performance rate of a variable is elaborated and its relation with hazard rate of industrial context and density function of the variable is studied. We prove characterization theorems for a general form of extended uniform distribution based on invariance of performance rate under scale transformations in a countable dense set. Applications of the distribution in quality control in industrial production, yield data of tuber crops among others are discussed.

Keywords Extended uniform distribution · Performance rate · Hazard rate
Cauchy equation · Tuber crop · Generalized extreme value distribution

1 Introduction

Modeling observed phenomena with theoretical distribution often requires an in-depth analysis of resultant data. Sometimes this indicates the possibility of a new model. Consider a nonnegative random variable X with unknown upper bound. If the variable $(X/\theta)^\alpha, \alpha > 0, \theta > 0$ follows an uniform distribution, then X is distributed as extended uniform, see Dasgupta (2017). Uniform distribution puts equal mass to intervals of constant length in its support, the variate values are homogeneously distributed in probability; whereas, in the above model a positive power and scale transformation of a nonhomogenous data attempts to achieve uniformity. A number of observed phenomena may be explained by this distribution.

R. Dasgupta (✉)
Theoretical Statistics and Mathematics Unit, Indian Statistical Institute,
203 B.T. Road, Kolkata 700108, India
e-mail: rdgupta@isical.ac.in; ratandasgupta@gmail.com

R. Dasgupta (ed.), *Advances in Growth Curve and Structural Equation Modeling*,
https://doi.org/10.1007/978-981-13-0980-9_3

Dasgupta (2017) introduced this distribution while modeling the number of cloves per garlic bulb. We studied the relationship of number of cloves x with observed cumulative frequency distribution $nF_n(x)$ in log-log scale. An approximate linear relationship suggests possibility of the following model for c.d.f.

$$F(x) = (x/\theta)^{\alpha}, \alpha > 0, x = 1, 2, 3, \ldots, \theta \qquad (1)$$

The new type of distribution with unknown upper bound θ and exponent $\alpha > 0$ is termed as discrete extended uniform distribution. The continuous version of the distribution is also studied therein. A number of characterization theorems were also proved.

The continuous version of the distribution in (1) may be written as

$$G(y) = (y/\theta)^{\alpha}, \alpha > 0, y \in (0, \theta] \qquad (2)$$

A scaled version of the variable with proper shaping turns out to be uniform over the range (0, 1]. As such we may term such a distribution as *extended uniform distribution*.

The above model resembles power law, but this has a *positive* exponent α; support of the variable has an unknown upper bound.

The maximum likelihood estimate of the parameters in (2) based on n iid observations $y_i, i = 1, 2, 3, \ldots, n$ are $\hat{\theta} = y_{(n)}$, the maximum observation; and $\hat{\alpha} = [-\frac{1}{n}\sum_{i=1}^{n}\log(\frac{y_i}{\theta})]^{-1}$.

A discretized version, the nearest integer $z = \lfloor y + \frac{1}{2} \rfloor$ of the continuous variable y within range $(0, \theta]$, with c.d.f F given in (2) is of interest.

In the present paper, apart from modeling yield data of tuber crops in agriculture, we examine different aspects of the distribution with applications in industrial context of reliability of a product. The concept of *performance rate* is also termed as reversed hazard rate, see e.g., Kijima (1998). We study its relation with hazard rate and density function of lifetime. A characterization of the distribution in terms of performance rate is proved.

In Sect. 2, we explain the concept of 'performance rate', compute the rates for generalized extreme value distributions, that include Weibull distribution. We also prove the characterization theorem for a class of distributions including extended uniform distribution based on invariance of performance rate on a countable dense set. Some examples of the variables following extended uniform distribution are also given in Sect. 3. Discussion of the results are made in the last section.

2 Performance Rate, Hazard Rate, a General Form of Extended Uniform, and Characterization Theorems

2.1 Relationship Between Performance Rate and Hazard Rate, Related Results

Performance rate of a distribution with density function $f(t)$ for lifetime T is defined as

$$s(t) = \lim_{\Delta t \to 0} P(t - \Delta t < T \le t | T \le t)/\Delta(t) = \frac{f(t)}{F(t)} \tag{3}$$

This can be interpreted as in terms of the limiting probability of a component failing just before taking record of its lifetime at t, given that the lifetime T is less than or equal to t, thus providing a measure of performance of a job under the given restriction on lifetime. The component has failed spending its life till time t, given that its lifetime is at most t, for high value of $s(t)$; thus, $s(t)(\ge f(t))$ may be termed as *performance rate* of the product. On the other hand, the previously used terminology 'reversed hazard rate' does not convey the inherent meaning of this measure, as explained in the present context.

The measure $s(t)$ relates to performance of the product around time t, when lifetime is conditioned to be within the time period $(0, t]$.

It is easy to see that for extended uniform distribution (2), $s(t) = \alpha/t$; a decreasing function of time t.

Performance rate $s(t)$ is related to hazard rate $r(t) = \frac{f(t)}{1-F(t)}$ and density function $f(t)$ of the lifetime T in the following manner.

$$\frac{1}{f(t)} = \frac{1}{r(t)} + \frac{1}{s(t)} \tag{4}$$

Thus, harmonic mean of performance rate and the hazard rate equals to $2f(t)$, twice the density function.

Let t_{med} be median of lifetime T. For $t \ge t_{\text{med}}$, $s(t) \le r(t)$; the hazard rate. For $t \le t_{\text{med}}$, inequality in other direction holds; $s(t) \ge r(t)$.

Performance rate characterizes a distribution from the relationship

$$\int_{\nu}^{x} s(t)dt = \int_{\nu}^{x} \frac{f(t)}{F(t)}dt = \log F(x) - \log F(\nu), \nu < x \tag{5}$$

where $F(\nu) > 0$. Thus, $F(x) = F(\nu)e^{\int_{\nu}^{x} s(t)dt}$.

A characterization result in terms of $s(t)$ has a possibility of reformulation in terms of $r(t)$, and vice versa.

A similar relationship in terms of hazard rate is $F(x) = 1 - e^{-\int_{-\infty}^{x} r(t)dt}$.

Performance rate is easy to compute for some distributions, whereas for some other distributions hazard rate takes a simpler form.

Some Examples.
Generalized extreme value distributions $GEV(\mu, \sigma, \xi)$ have applications in explaining limiting distribution of maximum of a sequence of random variables. Here, $\mu \in (-\infty, \infty)$ is the location parameter, $\sigma > 0$ is the scale parameter, and $\xi \in (-\infty, \infty)$ is the shape parameter, with support of the distribution as $[\mu - \sigma/\xi, \infty)$ for $\xi > 0$; $(-\infty, \infty)$ for $\xi = 0$, and the support is $(-\infty, \mu - \sigma/\xi]$ for $\xi < 0$.

A $GEV(\mu, \sigma, \xi)$ variable has distribution function $F(x) = \exp(-t(x))$, and density function $f(x) = \frac{1}{\sigma}t(x)^{\xi+1}\exp(-t(x))$, where $t(x) = e^{-(x-\mu)/\sigma}$ for $\xi = 0$, and $t(x) = (1 + \xi(\frac{x-\mu}{\sigma}))^{-1/\xi}$ for $\xi \neq 0$.

A very special case is Weibull distribution with applications in lifetime of industrial products. Characterization theorems for Weibull distribution, both discrete and continuous; with applications in modeling lifetime of Elephant foot yam plants are discussed in Dasgupta (2014).

Performance rate for a $GEV(\mu, \sigma, \xi)$ variable is then $s(x) = \frac{1}{\sigma}t(x)^{\xi+1}$.

2.2 *Characterization of Extended Uniform Distribution*

Let X be a random variable with support $(0, \theta]$; $\theta(> 1)$ having continuous density function f and cumulative distribution function F satisfying $F(u) = F(v)e^{\int_v^u s(y)dy}$, $u > v$ where $s(y) = \frac{f(y)}{F(y)}$ is the performance rate and $F(v) > 0$ for some $v > 0$. The assumption $\theta > 1$ is not restrictive, as a suitable scaled version of the variable will have upper bound exceeding 1, with the transformed variable inheriting similar properties.

One may define performance rate to be zero outside the support. We prove the following.

Theorem 1 *Let the performance rate s over the range $(0, \theta]$, $\theta > 1$ satisfies $s(cx) \propto s(x)$, $x = c^m$, $m \in N^+$, the set of positive integers and $(0 <)c \in A_0$, a countable dense neighborhood of 1 (e.g., $c \in A_0 = (1 - \epsilon, 1 + \epsilon) \cap Q$, $\epsilon > 0$, and Q is the set of rational numbers). Then, the variable X has distribution function F satisfying $F(u) = F(v)\exp(\int_v^u az^\beta dz)$, $u > v$, where $F(v) > 0$ for some $v > 0$. The variable is extended uniform iff the constant of proportionality in s is c^{-1}.*

Proof of the Theorem
One-way implication is clear, we prove the "only if" part. Write $s(c) = bs(1)$, where b is the constant of proportionality, $s(c^2) = bs(c) = b^2s(1)$, $s(c^3) = bs(c^2) = b^3s(1)$, and for integers m, $s(c^m) = b^m s(1)$. Write, $c^m = x$, i.e., $m = \log x/\log c$, then $s(x) = b^{\log x/\log c}s(1) = e^{\beta \log x}s(1) = ax^\beta$, $\beta = \log b/\log c$, $a = s(1)$.

This specifies the performance rate s and hence the distribution function F in a dense set $\{c, c^2, \ldots, c^m, \ldots\}$ of $(0, \infty)$, intersected with support of F, where s is nonzero. For an arbitrary real number $z > 0$, there exist integer m and $c \in A_0$, a dense set in $(1-\epsilon, 1+\epsilon)$; e.g., $c \in A_0 = (1-\epsilon, 1+\epsilon) \cap Q, \epsilon > 0$, such that c^m is arbitrary close to the number z. Next from continuity of f, the form of s is specified as $s = s(x) = ax^\beta$. The result is related to solution of Cauchy equation, see Dasgupta (2014).

Thus F is specified, $F(u) = F(v)e^{\int_v^u ax^\beta dx}, 0 < u \leq \theta$. Choice of v may be made arbitrary close to origin.

When the performance rate $s(x)$ is nonincreasing in x, $\beta \leq 0$.
A special case is $bc = 1$, then $\beta = -1$ and the distribution is extended uniform with $s(x) = a/x$.

Distribution of a random variable X with support in $(\theta_1, \theta_2], \theta_1 > 0$; such that $(\frac{X-\theta_1}{a})^\alpha$, $(\alpha > 0)$ is uniform over its permissible range may be termed as general extended uniform.

Characterization theorems for such distributions may be derived in a similar manner by translating the variable x to $x - \theta_1$.

3 Some Applications

3.1 Motor Failure Data

Distance in thousands of miles to successive major motor failures in buses operated by a large city bus company was analyzed in Davis (1952). The data on second bus failure had a poor fit in conventional models. In the data analysis therein, value of $\chi^2 = 16.57$ with $d.f. = 4$; for normal model, and $\chi^2 = 39.53$ with $d.f. = 6$; for exponential model, the values are high.

We now asses adequacy of extended uniform distribution in explaining abovementioned failure data. Table 1 provides the observed and expected frequencies under extended uniform model.

The last class interval of mileage was indeed an open interval, 160—up. We estimated the upper end point as 165 by plotting the values of $(x, nF_n(x))$; these should be along a straight line under the model (2), next varying the upper end point in order to maximize correlation coefficient among these points in a linear fit by least squares, and observing that maximum value $r = 0.9955$ is attained when upper value of x is taken as $x = 165$. See Fig. 1. Thus, we take the estimated value of a in model (2) as 165.

The value of α is estimated as the slope 0.83410 from least square fitted line. The m.l.e. from the grouped data is 0.856602. This is pretty close to the former estimate from regression shown in Fig. 1, having a good fit of least square line to the points $(x, nF_n(x))$.

A second estimate 156.4693974 of a is also available from the intercept 0.42980 of linear fit under model (2). This estimate can be compared with the former estimate 165, obtained by maximizing the value of correlation coefficient.

Table 1 Second motor failure of bus

Class interval	i	Frequency (Oi)	Cum frequency	Theoretical cum frequency	Theoretical frequency (ei)	Oi^2	Oi^2/ei
0–20	20	19	19	17.89036087	17.89036087	361	20.17846385
20–40	40	13	32	31.89394347	14.00358261	169	12.06834028
40–60	60	13	45	44.72866533	12.83472186	169	13.1674065
60–80	80	15	60	56.85875416	12.13008883	225	18.5489161
80–100	100	15	75	68.49044397	11.63168981	225	19.34370703
100–120	120	18	93	79.73978471	11.24934073	324	28.80168782
120–150	160	7	100	101.3646346	21.62484984	49	2.265911687
160–165	165	4	104	104	2.635365448	16	6.071264239
Total		104			104		120.4456975
Chi-square (1):	16.4457	d.f. = 5					
Chi-square (2):	10.72131	d.f. = 4					

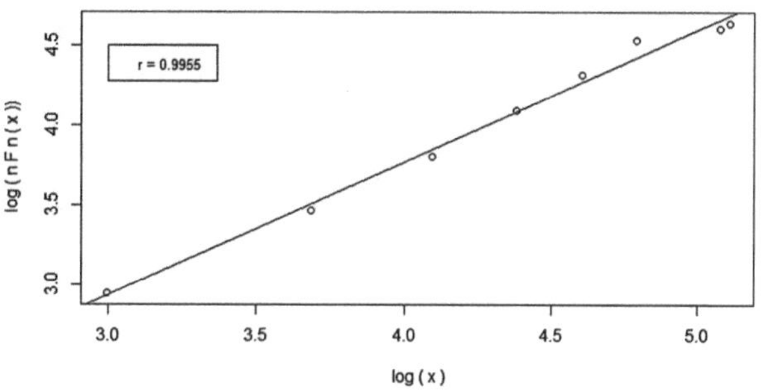

Fig. 1 Extended uniform model fit to second motor failure

In Table 1, the value of $\chi^2 = 16.4457$ with d.f. 5, p value is 0.0057 which is significant. However, merging the last two classes, the observed frequency is 13 and expected frequency is 22.187 in the merged class, and the resultant $\chi^2 = 10.72131$ is low with d.f. 4; p value is 0.0299 which is insignificant at 1% level.

The values of χ^2 reported in Davis (1952) in traditional model fit are higher than those obtained for present model (2).

3.2 *Repair Time for Air Conditioners*

Repair time of a faulty system depends on the extent of damage encountered in the system. Data in Table 2 relates to repair time in hours of 1203 air conditioning machines of food processing companies during the years 2011–12, reported in Burhanuddin et al. (2014).

To see the adequacy of the model (2) in repair time data, we plot the values of $(x, nF_n(x))$ in Fig. 2. The points should lie along a straight line under the model (2) in log-log scale, which is the case with a high value of correlation coefficient $r = 0.9980$, see Fig. 2.

Maximum likelihood estimator of the scale parameter a is the maximum observation in data, thus $a = 768$. The value of shape parameter α is estimated from the slope of least square regression line, thus $\alpha = 0.312693$. The m.l.e of α from grouped data is 0.294526, this is quite close to the former estimate. Figure 2 suggests that we may rely on the former estimate.

A second estimate $a = 740.4058$ is also available from the intercept 5.026549 of linear fit under the model (2). This estimate may be compared with the m.l.e. $a = 768$.

In Table 2, the value of $\chi^2 = 23.19547$ with d.f. 13, the p value is 0.03939.

This is significant at 5% level but not significant at 1% level. Since $p \approx 0.04$ and a there is a close fit for least squared line to the points $(x, nF_n(x))$ in log-log scale with $r = 0.9980$; we may consider (2) to be a satisfactory model for repair time.

Table 2 Repair time (hour)

Class interval (hour)	i	Frequency (Oi)	Cum frequency	Theoretical cum frequency	Theoretical frequency (ei)	Oi^2	Oi^2/ei
0–48	43	530	530	505.5286082	505.5286082	280900	555.6559915
49–96	96	92	622	627.8786416	122.3500335	8464	69.1785671
97–144	144	80	702	712.7516217	84.87298005	6400	75.40680198
145–192	192	70	772	779.8403142	67.0886925	4900	73.03764341
193–240	240	65	337	836.1973082	56.356994	4225	74.96851234
241–288	288	63	900	885.2545886	49.05728041	3969	80.90542253
289–336	336	41	941	928.97062	43.71603139	1681	38.4527128
337–384	334	47	988	968.5803518	39.60973185	2209	55.76912281
385–432	432	42	1030	1004.918151	36.33779919	1764	48.54449194
433–480	480	32	1062	1038.577088	33.65893742	1024	30.4228261
481–528	528	29	1091	1069.995474	31.41838517	841	26.76776656
529–576	576	28	1119	1099.507406	29.5119323	784	26.56552583
577–624	624	29	1148	1127.37403	27.86662408	841	30.17947196
625–672	672	21	1169	1153.803764	26.42973416	441	16.6857524
673–720	720	16	1185	1178.965894	25.1621298	256	10.17401953
721–768	768	18	1203	1203	24.03410603	324	13.48084258
Total		1203			1203		1226.195471
Chi-square =	23.19547133	d.f. = 13					

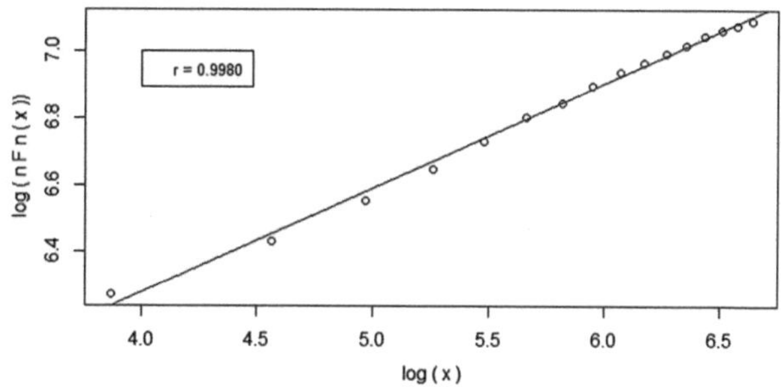

Fig. 2 Extended uniform model fit for repair time (hour)

3.3 Number of Elephant Foot Yam Tubers per Pit

Elephant foot yam is a staple food and cultivated as a cash crop. Yam of average seed corm weight 500 g was planted in 100 pits in a growth experiment conducted during the years 2016–17 in the farmland of Indian Statistical Institute, Giridih. The experimental plot consisted of loamy soil that is conducive of size expansion for underground yam in the period of growth. As a result, there were up to four distinct tubers in a single pit out of a seed corm. One pit did not produce any yam. In some pits only trace of yam growth were seen, these are counted as a single tuber.

We fit the discrete model (1) for number of tubers given in Table 3. Approximate linear relationship of the points $(x, nF_n(x))$ in log-log scale with $r = 0.9765$ in Fig. 3 is an indication of model appropriateness.

Maximum likelihood estimator of the scale parameter a is the maximum observation in data, thus $a = 4$. Although there is a single observation with tuber no. as 4, the expected frequency in that class under the model is greater than 4; see Table 3. This validates chi-square test.

The value of shape parameter α is estimated from the slope of least square regression line, thus $\alpha = 0.15005$.

Intercept of the least square fitted line provides a second estimate of scale parameter as $a = 3.490049$, which is pretty close to the value 4 of m.l.e. This is another indication of model adequacy. The value of χ^2 in goodness of fit, with d.f. 1 is 3.6419, which is insignificant with p value 0.05634.

3.4 Canopy Radius of Elephant Foot Yam Plants

Canopy radius measures spread of canopy in yam plants absorbing sunlight. The maximum length of three leaf segments in a nearly circular structure that usually appear on the top of a stem is considered for canopy radius measurement in that stem. The maximum reading over all stems in the plant sprouting from a pit is taken as 'canopy radius' for the plant.

Table 3 No of yam tubers per pit in 99 pits

X	Frequency (Oi)	Cum frequency	Theoretical cum frequency	Theoretical frequency (ei)	Oi^2	Oi^2/ei
1	81	81	80.40741363	80.40741363	6561	81.59695361
2	12	93	89.22070359	8.813289962	144	16.33896089
3	5	98	94.81742071	5.596717121	25	4.46690434
4	1	99	99	4.182579288	1	0.23908692
Total	99					102.6419058
Chi-square =	3.641905762	d.f. = 1				

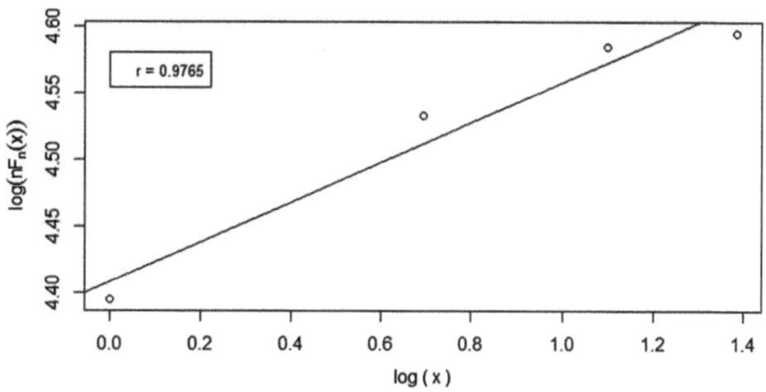

Fig. 3 Model fit for no. of tubers per pit in yam plants (2016–17)

In this growth experiment sprouting started from May 2, 2016, and the experiment continued till the harvesting on December 28, 2016 at Indian Statistical Institute, Giridih Farm. The average seed weight of yam is 500 g. We consider canopy radius recorded on August 28, 2016; the time around middle of the experiment. The number of stems per plant varied from 1–5.

Canopy radius is an important indicator of plant photosynthesis that affects underground yam deposition. We wish to model this variable by extended uniform distribution.

Approximate linear relationship of 97 points $(x, nF_n(x))$ on canopy radius x and number of observations falling within that reading from below $nF_n(x)$ in log-log scale deleting two topmost radius outliers 74 and 74.5 cm. are shown in Fig. 4 in log-log scale that shows $r = 0.9971$, indicating a fair possibility of adequate modeling of canopy radius by extended uniform distribution.

The m.l.e. of the scale parameter a is the maximum observation excluding the outliers, thus $a = 71$.

The value of shape parameter α from the slope of least square regression line is $\alpha = 2.71879$.

Intercept of the least square fitted line provides a second estimate of scale parameter as $a = 66.69$, which may be compared with the m.l.e. 71.

To see the model adequacy we now group the data as shown in Table 4. The m.l.e. of the shape parameter α from the group data is $\alpha = 2.66853$. The value is pretty close to the other estimate of α as 2.71879, which is the slope of least square regression line in raw data shown in Fig. 4.

Chi-square goodness of fit with 12 d.f. indicates that the model is appropriate, $\chi^2 = 19.9616$ when m.l.e. of α in grouped data is considered, and $\chi^2 = 20.4788$ when intercept estimate of α in raw data is considered. Respective p values are 0.0678 and 0.0586; and these are insignificant.

Table 4 Canopy radius of 97 yam planets (2016)

Class interval	Frequency (Oi)	Cum frequency	Theoretical cum frequency	Theoretical frequency (ei)	Oi^2/ei
(0–24]	6	6	5.367454324	5.367454824	6.70708952
(25.5–30]	6	12	9.735890368	4.368435544	8.240936518
(33–36.5]	6	IS	16.43079853	6.694908165	5.377220884
(37–42]	9	27	23.89586312	7.465064586	10.85054242
(42.5–44.5]	6	33	27.88248216	3.986619045	9.030208202
(45–47]	6	39	32.26090823	4.378426068	8.222132666
(48–51]	7	46	40.11765461	7.856746374	6.236678349
(51.2–52]	6	52	42.25126299	2.133608382	16.8728246
(52.5–53.2]	6	53	44.903507	2.652244013	13.57341173
(53.5–56]	7	65	51.49030788	6.586800881	7.43911967
(56.2–53]	6	71	56.54494639	5.054638504	7.122171046
(59.5–60.5]	6	77	63.285026	6.740079617	5.341183198
(62–64]	6	S3	73.53239965	10.24737365	3.513095281
(64.5–67.5]	6	89	84.75889572	11.22649607	3.20669956
(68.2–71]	3	99	97	12.24110428	5.228286479
Total	97			97	116.9616001
Chi-square =	116.9616001-97	= 19.9616	d.f. = 12		

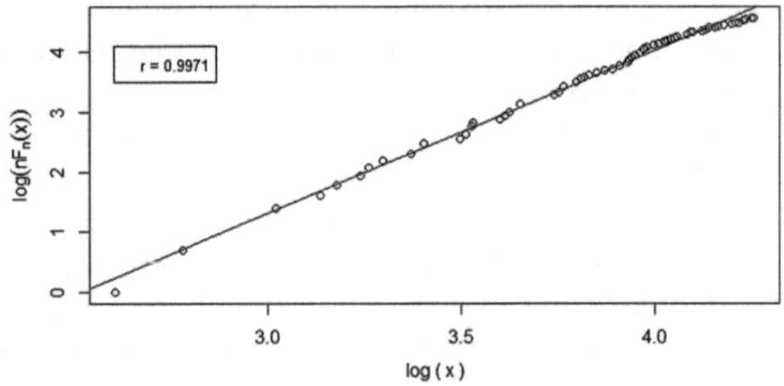

Fig. 4 Model fit for canopy radius of yam plant (2016)

4 Discussions

Extended uniform distribution is a variant of uniform distribution. Uniformity in data is achieved by scale and shape transformation. In its more general form, a location change may also be considered to achieve uniformity. The distribution may be characterized by the inversely proportional performance rate $s(x) = a/x$ in thesupport of the variable. A number of data sets, discrete and continuous may be explained by this distribution as seen in the given examples, where traditional models fail.

Performance rate characterizes a distribution, when value of the c.d.f. at some lower point is specified. The performance rate for a generalized extreme value distribution, $GEV(\mu, \sigma, \xi)$ is $s(x) = \frac{1}{\sigma}t(x)^{\xi+1}$. By a transformation of the variable X to $t(X)$, a $GEV(\mu, \sigma, \xi)$ variable may be changed to an extended uniform variable when $\xi = -2$.

References

Burhanuddin, M. A., Ghani, M. K. A., Ahmad, A., Abas, Z. A., & Izzah, Z. (2014). Reliability analysis of the failure data in industrial repairable systems due to equipment risk factors. *Applied Mathematical Sciences*, 8(31), 1543–1555. https://doi.org/10.12988/ams.2014.4278.

Dasgupta R. (2014). Characterization theorems for weibull distribution with applications. *Journal of Environmental Statistics*. (UCLA, Dept. of Stat.), *6*(4), 1–25.

Dasgupta, R. (2017). Model selection and validation in agricultural context: Extended uniform distribution and some characterization theorems. *Growth Curve Models and Applications*. Chapter 9. Springer.

Davis, D. J. (1952). An analysis of some failure data. *Journal of the American Statistical Association*, *47*(258), 113–150.

Kijima, M. (1998). Hazard rate and reversed hazard rate monotonicities in continuous-time Markov chains. *Journal of Applied Probability*, *35*, 545–556.

Do We Share the Same Destiny? An Empirical Study on the Convergence of Growth Among Indian States

Sattwik Santra and Samarjit Das

Abstract The paper examines whether there is a convergence of per capita consumption, and state domestic product across the various states of India. We use a set of panel unit root tests including ones that are robust to cross-sectional dependence. We find that although there is some evidence in favor of convergence in per capita consumption, the state domestic products in per capita terms of the Indian states exhibit non-convergence for all the tests. But there are certain state groups that exhibit intra-group convergence for all the variables considered for all the tests. The results indicate exacerbating overall inter-state inequality prospects on one hand, and a bunching of the states in terms of the overall economic performance on the other.

Keywords Growth convergence · Empirical methodology · Panel unit root · Cross-sectional dependence

1 Introduction

In the recent empirical literature on neoclassical growth, the issue of convergence is one of the key topics of interest to economists dealing with economic development and policymakers, in developing countries. Convergence is defined as the phenomenon where interregional disparity in a particular variable tends to ease out over time and the regions become more and more homogeneous with respect to that particular attribute. Neoclassical growth theory ascribes the mobility of supply-side

---

The authors are grateful to an anonymous referee for insightful and constructive comments on an earlier draft of this paper.

S. Santra
Centre for Training and Research in Public Finance and Policy, Centre for Studies in Social Sciences, R-1, B.P. Township, Kolkata 700094, India

S. Das (✉)
Economic Research Unit, Indian Statistical Institute, 203, B.T. Road, Kolkata 700108, India
e-mail: samarjit@isical.ac.in

R. Dasgupta (ed.), *Advances in Growth Curve and Structural Equation Modeling*,
https://doi.org/10.1007/978-981-13-0980-9_4

factors particularly capital, labor and technology, and trade in commodities as possible reasons for the eventual reduction of interregional disparities. In a country, the relative ease of factor mobility and commodity trade is purported to relieve interregional disparity. However, the presence of geographic, demographic, and economic diversity may serve as an impediment to interregional convergence—at least in the short run. If such differences are allowed to accumulate over time, they may eventually compromise the overall well-being (Alesina and Perotti 1996; Benhabib and Rustichini 1996; Chowdhury 2003). To serve an example, regional problems in China—an obstacle to its stability and development, are primarily attributed by Wei and Kim (2002) to increasing regional inequality. In the present study, we analyze the nature of convergence in the context of India. Specifically, we study the convergence of some major macroeconomic aggregates across the states of India. It is a fact that India is quite a vast country and the different regions have different resource bases, along with rigidities in factor mobility added to a huge sociocultural diversity. Thus, it is an interesting exercise to analyze the depth and the extent of this heterogeneity in the context of convergence.

Following the seminal work of Baumol (1986), Barro and Sala-i-Martin (1992, 1995) introduced two notions of convergence. The first of these notions is termed as β-convergence (beta convergence) is used to assess whether the poor states grow at a relatively higher rate than the richer ones. If it is indeed the case, then it implies that over time, the poor states are able to catch up with the richer states. Alternatively, convergence may also be viewed as a reduction in income inequality. This constitutes the second notion of convergence which is called σ-convergence (sigma convergence). The literature on empirical estimation of convergence produced a number of articles. Based on cross-section data for 100 years, Barro and Sala-i-Martin (1992) found evidence of unconditional convergence among the U.S. states. The use of cross-section data was criticized by Bernard and Durlauf (1995) and Evans and Karras (1996) and advocated the use of time series unit root tests to test for convergence. The econometric literature has provided justifications that invalidate the use of traditional growth regressions for studying economic convergence along the lines of popular notions of β and σ convergence. In particular, these methods fail to allow for unobserved (and persistent) heterogeneity across the regions. In the presence of these unobserved regional heterogeneities, any unit root test that omit these fixed effects may render such test statistics inconsistent as it leads to the problem of endogeneity.

In the literature on panel data framework, the approach to test for convergence involves conducting a panel unit root test of the relevant variable/s relative to that of a common numeraire. If the unit root test accepts the hypothesis of stationarity of the variable/s with respect to the common numeraire (which is often taken to be some weighted average of the variable/s over the cross sections), then it indicates the mean-reverting nature of the variable/s and any deviations from the numeraire are deemed to diminish over time. This suggests the temporary nature of the deviations from the numeraire and one may conclude that convergence holds.

In the context of panel data models, a problem that needs to be addressed is the problem of spatial autocorrelation. Panel data across regions are expected to be spatially correlated. A number of factors attribute to such dependency and these include the very reasons that are responsible for convergence (which includes regional factor mobility and commodity trade. See also Temple 1999). Apart from these factors, the presence of a single currency, common policies adopted by the central government, etc., also contribute toward spatial dependence. Most importantly, in the context of panel data framework, since the very construction of the tests for convergence involves the use of a common numeraire, renders the data to be cross-sectional dependent. This follows since such construction involves the use of a ratio that have a common divisor (see Pearson (1897) also Kim 1999). In such circumstances, the standard errors and tests that do not account for these correlations yield estimates that are misleading as in Evans and Karras (1996).

The effect of contemporaneous correlation on standard panel unit root tests, such as those proposed by Maddala and Wu (1999), Levin et al. (2002), and Im et al. (2003) are presented by O'Connell (1998) and Breitung and Das (2005). These later authors have highlighted that in the presence of contemporaneous correlation, the standard tests are more likely to reject the null hypothesis of a presence of unit roots. In this study, we present results that arise from tests suggested by Levin et al. (2002), and Im et al. (2003) and present it alongside Breitung and Das (2005) for comparison.

The literature on convergence in the context of India is relatively scarce. The paper by Nagaraj et al. (1997) considered the growth performance of Indian states during the 1960–1994 period and found evidence of conditional convergence. Using a cross-sectional data, Bajpai and Sachs (1996) considered the unconditional β-convergence during 1960–1970 and found statistically significant evidence in its favor within this time period but not thereafter. Cashin and Sahay (1996) using the same dataset, studied the extent of β-convergence but failed to find statistically significant rate of convergence during the period 1960–1992. Contrary to these findings, a number of studies have also reported a statistically significant divergence in growth patterns across the Indian states. Examples include Raman (1996), Marjit and Mitra (1996), Dasgupta et al. (2000) among others. Our present study adds on to these contributions and provides an updated testimony on the topic.

The plan of this article is as follows. The next section briefly discusses the methodology. The following section presents the empirical results based on panel unit root tests and this is followed by forwarding the conclusions.

2 Methodology

Here, we provide an introduction to the econometric methodology of panel data models used to test for the presence of unit roots. Note that this forms the basis of the tests used to ascertain convergence. We consider the data generating process:

$$\Delta y_{it} = \alpha_i + \rho y_{it-1} + \sum_{j=1}^{p} \gamma_{ij} \Delta y_{it-j} + \varepsilon_{it}$$

Here y is the variable of concern, t indexes time, and i indexes the cross-section units, α represents the unobserved fixed effects and the errors (ε_{it}) are independent and identically distributed over time with $E(\varepsilon_t \varepsilon_t') = \Omega$, where Ω is not necessarily a diagonal matrix. The null of a unit root posits that $\rho = 0$ that is, all the time series are random walks. Under the alternative of stationarity, it is assumed that $\rho < 0$. If the variable under scrutiny exhibits stationarity, then it implies regional convergence with respect to that variable. Typically, the test involves considering the variable in logarithms and demeaned from the common or average trend which constitutes our numeraire. For our purpose, we consider the simple cross-section mean of the relevant variable considered in logarithms as our numeraire. To pacify concerns about the choice of numeraire, it needs to be mentioned at the very outset that we have tested various specifications about the choice of numeraire and found that these yield similar conclusions. As a technical note, it needs to be mentioned that since we allow for the possibility of nonzero region-specific intercepts in above equation, the nature of convergence the is captured by the above model is a conditional one. This is opposed to the notion of absolute convergence (see Evans and Karras 1996).

If the null hypothesis of a unit root is rejected, then the variable in question is mean reverting implying that any deviations from the numeraire is temporary in nature and should decline over time. This upholds the claim of regional convergence of the variable. Instead, if the unit root tests fail to reject the null hypothesis, then it indicates that the variable follows a random walk and hence, the deviation from the numeraire is persistent over time indicative of divergence of the variable.

3 Data Description

The study examines the convergence of two major macroeconomics aggregates across the Indian states. The first of these two variables is the state per capita domestic product and the second is the aggregate monthly per capita consumption expenditure (MPCE) of the states. For the state domestic product, we consider both the gross and the net state per capita domestic products (GSDP and NSDP respectively). State domestic product is a measure of the aggregate output arising out of the economic activity within the geographical boundary of the state. Hence, convergence in the per capita state domestic product would imply that the aggregate per capita productions of the concerned states are approaching a common trend. This is more in tune with the standard growth theoretic models which discuss the inter-temporal trajectory of the aggregate output. In the context of convergence, a yet another variable of interest might be the per capita consumption expenditures of the states. This particular variable, unlike the per capita state domestic product, is directly related to the well-being of the individuals of the concerned state and reflects the average level of utility of the household in the states. As a side note, it is worthwhile to mention that the use of consumption expenditure has become the present practice in the poverty measurement literature (Lipton and Ravallion 1985; Lanjouw and Lanjouw 2001; Alkire and Foster 2011, etc., among others). There is a caveat though: individuals'

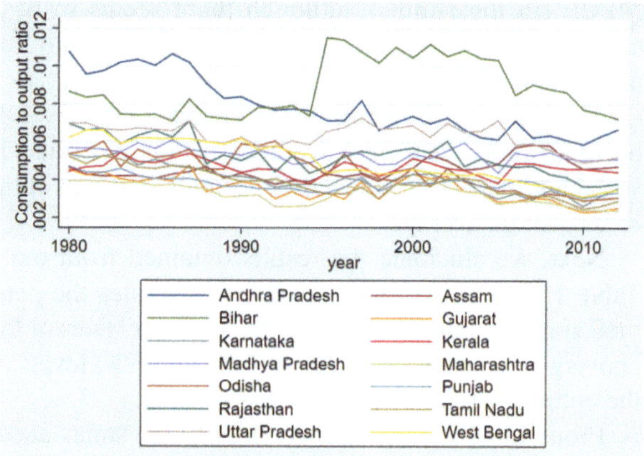

Fig. 1 Ratio of average per capita consumption expenditure of the states to the respective per capita state domestic products (*Source* The authors' calculation from data)

consumption patterns exhibit inter-temporal smoothing. As a result, if there is a regional divergence of aggregate output it may not get reflected if we consider per capita average consumption of the state. On the other hand, a divergence in the consumption patterns would indicate a rise in heterogeneity in individuals' tastes and preference characteristics. Figure 1 illustrates the time trend of the ratio between the average per capita consumption expenditure of the states to the respective per capita state domestic products. The figure exhibits that the ratio in general has a declining trend and exhibit some heterogeneity across the states.

For all studies, we consider 14 major states in India, viz., Andhra Pradesh, Assam, Bihar, Gujarat, Karnataka, Kerala, Madhya Pradesh, Maharashtra, Odhisha, Punjab, Rajasthan, Tamil Nadu, Uttar Pradesh, and West Bengal. The state domestic product series have been compiled from the data provided by the Ministry of Statistics and Program Implementation, Government of India. The state domestic product series are at a factor cost with constant 2004–2005 prices. MPCE of the states has been constructed from the 34 rounds of National Sample Survey data provided by National Sample Survey Organization, Government of India. To make the consumption expenditure comparable over time, we use the consumer price indices from Labour Bureau, Government of India. For urban consumption, we use the consumer price index for industrial worker, and for rural sector, we use the consumer price index for the agricultural labor. Population data have been obtained from the Economic Survey, 2014–2015, Government of India.

4 Results and Discussions

From an inspection of the aggregate per capita GSDP and NSDP (in logarithms) of the states presented in Figs. 2 and 3, it can be seen that over the years, there has been a persistence in the divergence among the states in levels. For the states'

MPCE (in logarithms), although there seems to be an initial convergence in the consumption patterns, from 1980 onwards, the plot registers a steady divergence in the consumption figures. This is illustrated in Fig. 4.

This divergence in both the per capita GSDP/NSDP and MPCE across the states becomes more clear once we inspect Figs. 5, 6, and 7. These figures plot the same variables, respectively, but after normalizing each of the variables with respect to the cross-section average.

Next, we elucidate the results obtained from our empirical model provided in Table 1 through to Table 5. Table 1 tabulates the panel unit root test statistics and their statistical significance for the 14 major states of India. The p-values indicate that convergence is achieved at significant (at 5%) levels only for MPCE if we consider the entire time horizon from 1958 to 2012.

From Table 2 onwards, we classify the states according to certain criterion and check for convergence. In Table 2, we consider only those states that are relatively affluent economically namely the states of Andhra Pradesh, Gujarat, Karnataka,

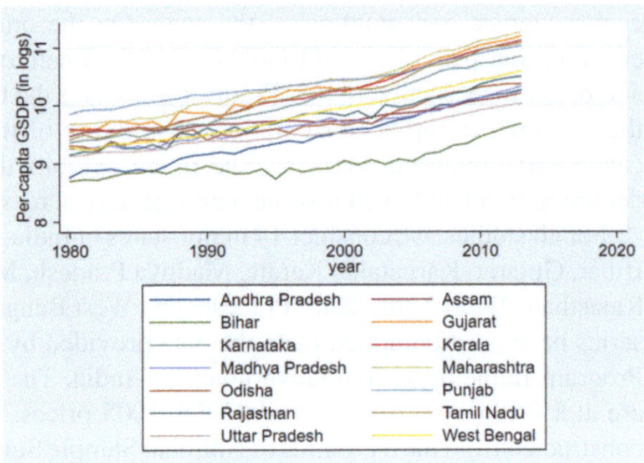

Fig. 2 Per capita gross state domestic products (*Source* Ministry of Statistics and Program Implementation, Government of India)

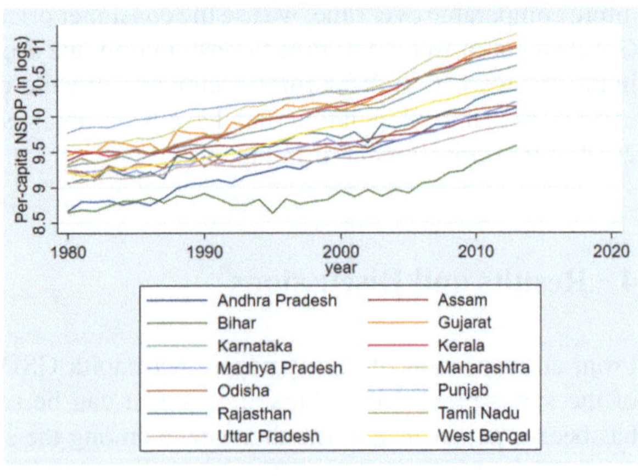

Fig. 3 Per capita net state domestic products (*Source* Ministry of Statistics and Program Implementation, Government of India)

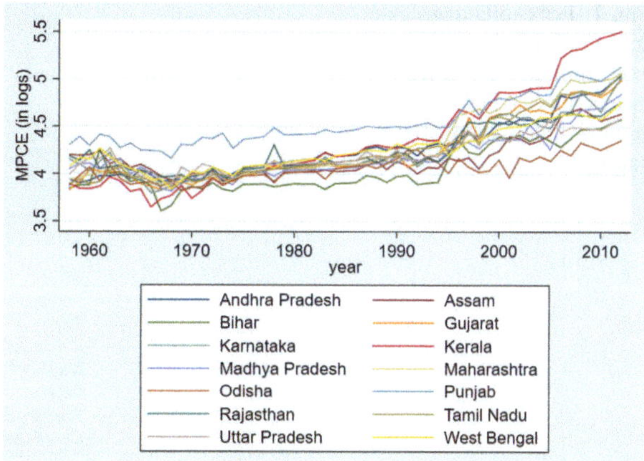

Fig. 4 Per capita aggregate state consumption expenditures (*Source* Ministry of Statistics and Program Implementation, Government of India)

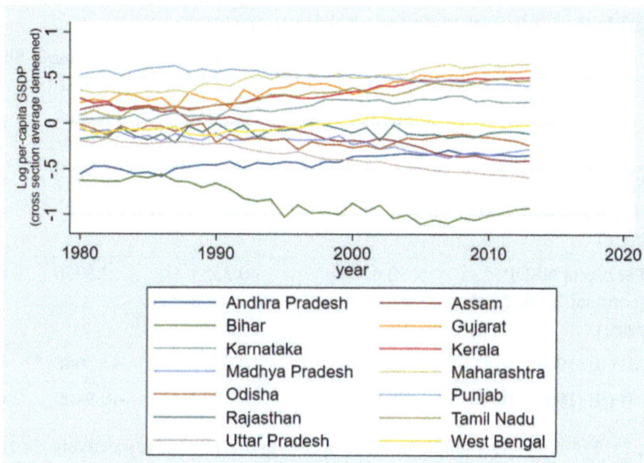

Fig. 5 Per capita gross state domestic products—cross section means removed (*Source* The authors' calculation from data)

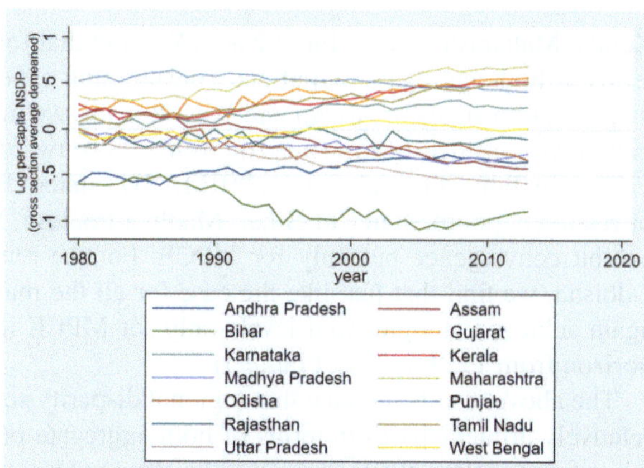

Fig. 6 Per capita net state domestic products—cross section means removed (*Source* The authors' calculation from data)

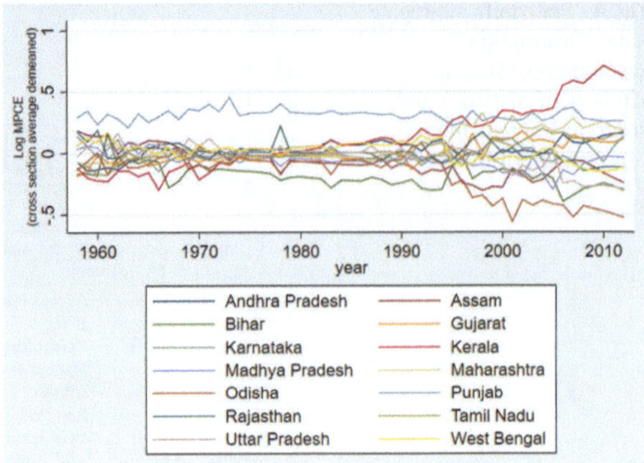

Fig. 7 Per capita aggregate state consumption expenditures—cross section means removed (*Source* The authors' calculation from data)

Table 1 Unit root tests for all 14 major states

	Levin-Lin-Chu unit root test		Im-Pesaran-Shin unit root test		Breitung unit root test	
	Statistic (Adjusted t*)	p-value	Statistic (W-t-bar)	p-value	Statistic (lambda*)	p-value
Per capita GSDP (constant 2004–2005 price)	0.5470	0.7078	2.8787	0.9980	0.3734	0.6456
Per capita NSDP (constant 2004–2005 price)	0.6289	0.7353	2.8776	0.9980	0.1027	0.5409
MPCE (1958–2012)	−1.7916**	0.0366	−3.3648***	0.0004	−2.3116**	0.0104
MPCE (1980–2012)	−0.9032	0.1832	−0.9409	0.1734	−1.3005*	0.0967

*, **, *** indicates significance at 10%, 5% and 1% levels respectively

Kerala, Maharashtra, and Tamil Nadu. We find that for these states, convergence in terms of both aggregate output and consumption expenditures is achieved. For the major eastern states of Assam and West Bengal, we find convergence in MPCE but only if we consider the time horizon from 1980 onwards. This is reported in Table 3. Figures in Table 4 reveal that the "BIMARU" states (Bose 2010, p. 64) an acronym of relatively poorer states of Bihar, Madhya Pradesh, Rajasthan and Uttar Pradesh, exhibit convergence but only for MPCE. For the remaining states of Punjab and Odhisha, we find that just like the case for all the major states, convergence is yet again achieved at significant levels only for MPCE if we consider the entire time horizon from 1958 to 2012 (Table 5).

The above results indicate that regional disparity seems to be diminishing for the relatively affluent states in terms of both aggregate output and consumption levels while for the relatively poorer states the same holds but only for consumption levels. The other states are subject to increasing disparities that are expected to accumulate over time, i.e., in the long run. Also, the gap between the rich and the poorer states in

Table 2 Unit root tests for the developed states of Andhra Pradesh, Gujarat, Karnataka, Kerala, Maharashtra, and Tamil Nadu

	Levin-Lin-Chu unit root test		Im-Pesaran-Shin unit root test		Breitung unit root test	
	Statistic (Adjusted t*)	p-value	Statistic (W-t-bar)	p-value	Statistic (lambda*)	p-value
Per capita GSDP (constant 2004–2005 price)	−4.1077***	0.0000	−4.2833***	0.0000	−4.0105***	0.0000
Per capita NSDP (constant 2004–05 price)	−3.7770***	0.0001	−4.0346***	0.0000	−4.2701***	0.0000
MPCE (1958–2012)	0.1321	0.5526	−0.9980	0.1591	−1.6737**	0.0471
MPCE (1980–2012)	−0.7039	0.2407	−0.7433	0.2286	−1.9942**	0.0231

*, **, *** indicates significance at 10%, 5% and 1% levels respectively

Table 3 Unit root states for the Eastern states of Assam and West Bengal

	Levin-Lin-Chu unit root test		Im-Pesaran-Shin unit root test		Breitung unit root test	
	Statistic (Adjusted t*)	p-value	Statistic (W-t-bar)	p-value	Statistic (lambda*)	p-value
Per capita GSDP (constant 2004–2005 price)	0.8576	0.8044	2.4748	0.9933	1.9436	0.9740
Per capita NSDP (constant 2004–2005 price)	1.0249	0.8473	2.6179	0.9956	1.8745	0.9696
MPCE (1958–2012)	−3.0672***	0.0011	−2.5183***	0.0059	−0.8980	0.1846
MPCE (1980–2012)	−2.1904**	0.0142	−1.7674**	0.0386	−2.5224***	0.0058

*, **, *** indicates significance at 10%, 5% and 1% levels respectively

Table 4 Unit root states for the BIMARU states, viz., Bihar, Madhya Pradesh, Rajasthan and Uttar Pradesh

	Levin-Lin-Chu unit root test		Im-Pesaran-Shin unit root test		Breitung unit root test	
	Statistic (Adjusted t*)	p-value	Statistic (W-t-bar)	p-value	Statistic (lambda*)	p-value
Per capita GSDP (constant 2004–2005 price)	0.9089	0.8183	1.5690	0.9417	−0.9251	0.1775
Per capita NSDP (constant 2004–2005 price)	0.9161	0.8202	1.3568	0.9126	−1.0373	0.1498
MPCE (1958–2012)	−4.9900***	0.0000	−5.5965***	0.0000	−4.4567***	0.0000
MPCE (1980–2012)	−2.1325**	0.0165	−1.7844**	0.0372	−3.5611***	0.0002

*, **, *** indicates significance at 10%, 5% and 1% levels respectively

Table 5 Unit root tests for the remaining states of Punjab and Odhisha

	Levin-Lin-Chu unit root test		Im-Pesaran-Shin unit root test		Breitung unit root test	
	Statistic (Adjusted t*)	p-value	Statistic (W-t-bar)	p-value	Statistic (lambda*)	p-value
Per capita GSDP (constant 2004–2005 price)	−2.5861***	0.0049	−2.5965***	0.0047	−1.2253	0.1102
Per capita NSDP (constant 2004–2005 price)	−3.3583***	0.0004	−3.3188***	0.0005	−0.8699	0.1922
MPCE (1958–2012)	0.8573	0.8044	1.2257	0.8898	−1.9460**	0.0258
MPCE (1980–2012)	0.4496	0.6735	1.3845	0.9169	−0.5940	0.2763

*, **, *** indicates significance at 10%, 5% and 1% levels respectively

terms of these aggregates seems to worsen with time: something that calls for serious policy intervention.

As some minor closing remarks, it may be noted that while for a majority of the cases the unit root tests bear similar results in terms of acceptance of the convergence hypothesis across the model that allow cross-sectional dependence and the ones that do not, there are some cases where the conclusions do not match up. Also, a minor observation is the fact that there seems to be no difference in conclusions drawn if we compare the results based on per capita GSDP to those based on per capita NSDP values.

5 Concluding Remarks

The present study looks at the nature of convergence of certain macroeconomic aggregates across the major states of India. The indicators looked at are the per capita gross state domestic product, the per capita net state domestic product and the states' aggregate per capita monthly consumption expenditures. Data on the states' domestic products covers the time span from 1980 to 2013, while the monthly expenditure figures date from 1958 to 2012. The check for convergence employs panel unit root tests inclusive of one that is robust to cross-sectional dependence. This methodology circumvents problems associated with traditional approaches.

Results indicate that convergence in aggregate output is not attained if we consider all the major states of India. But if we consider the relatively affluent states, then convergence hypothesis is found to be statistically significant. In terms of aggregate expenditure, the situation is altogether different. For this variable, we find a rather statistically weak convergence for all the major states taken collectively. Among the state groups, convergence in consumption holds for the relatively affluent states as well as the relatively poor BIMARU states.

In terms of policy prescriptions, these results indicate the need for a strong centralized action aimed toward the relatively economically backward states. These might take the form of redistributive policies such as private and public investments, regulation of FDI investments, credit disbursement by banks, central assistance, and policies aimed at providing factor mobility especially labor and human capital and ease of conducting trade. The present step by the government to bring in uniformity in the tax structure through the introduction of GST might aid in this process.

References

Alesina, A., & Perotti, R. (1996). Income distribution, political instability and investment. *European Economic Review, 81*(5), 1170–1189.

Alkire, S., & Foster, J. (2011). Counting and multidimensional poverty measurement. *Journal of Public Economics, 95*(7), 476–487.

Bajpai, N., & Sachs, J. (1996). *Trends in inter-state inequalities of income in India*. Development Discussion Paper No. 528, HID, Harvard University.

Barro, R., & Sala-i-Martin, X. (1992). Convergence. *Journal of Political Economy, 100*(20), 223–251.

Barro, R., & Sala-i-Martin, X. (1995). *Economic Growth*. New York: McGraw-Hill.

Baumol, W. (1986). Productivity growth, convergence and welfare: what the long-run data show. *American Economic Review, 76*(5), 1075–1085.

Benhabib, J., & Rustichini, A. (1996). Social conflict and growth. *Journal of Economic Growth, 1*(1), 129–146.

Bernard, A. B., & Durlauf, S. N. (1995). Convergence of international output. *Journal of Applied Econometrics, 10,* 97–108.

Bose, A. (2010). *Headcount: Memoirs of a demographer*. India: Penguin Books.

Breitung, J., & Das, S. (2005). Panel unit root tests under cross sectional dependence. *Statistica Neerlandica, 52,* 1–20.

Cashin, P., & Sahay, R. (1996). Internal migration, centre-state grants, and economic growth in the states of India. *IMF Staff Paper, 43*(1), 123–171.

Chowdhury, K. (2003). Empirics for world income distribution: What does the World Bank data reveal? *Journal of Development Areas, 36*(2), 59–83.

Dasgupta, D., Maiti, P., Mukherjee, R., Sarkar, S., & Chakrabarty, S. (2000). Growth and Interstate Disparities in India. *Economic and Political Weekly*, July 1, 2413–2422.

Evans, P., & Karras, G. (1996). Convergence revisited. *Journal of Monetary Economics, 37,* 249–265.

Im, K. S., Pesaran, M. H., & Shin, S. (2003). Testing for unit roots in heterogeneous panels. *Journal of Econometrics, 115,* 53–74.

Kim, J. H. (1999). Spurious correlation between ratios with a common divisor. *Statistics and Probability Letters, 44,* 383–386.

Lanjouw, J., & Lanjouw, P. (2001). How to compare apples and oranges: Poverty measurement based on different definitions of consumption. *Review of Income and Wealth, 47*(1), 25–42.

Levin, A., Lin, C., & Chu, C. J. (2002). Unit root tests in panel data: Asymptotic and finite sample properties. *Journal of Econometrics, 108,* 1–24.

Lipton, M., & Ravallion, M. (1995). Poverty and Policy. In J. Behrman & T. N. Srinivasan (Eds.), *Handbook of development economics* (Vol. 3B, pp. 2551–2557). Amsterdam: North-Holland.

Maddala, G. S., & Wu, S. (1999). A comparative study of unit root tests with panel data and a new simple test. *Oxford Bulletin of Economics and Statistics, 61,* 631–652.

Marjit, S., & Mitra S. (1996). Convergence in regional growth rates: Indian research agenda. *Economic and Political Weekly*, August 17, 31, 33, 2239–2242.

Nagaraj, R., Varoudakis, A., & Veganzones, M. A. (1997). *Long-run growth trends and convergence across Indian states*. Mumbai: Mimeo, IGIDR.

O'Connell, P. (1998). The overvaluation of purchasing power parity. *Journal of International Economics, 44,* 1–19.

Pearson, K. (1897). On a form of spurious correlation which may arise when indices are used in the measurement of organs. *Proceedings of Royal Society of London Series A, 60,* 489–498.

Raman, J. (1996). *Convergence or uneven development: a note on regional development in India*. Indiana: Mimeo, Valparaiso University.

Temple, J. (1999). The new growth evidence. *Journal of Economic Literature, 37,* 112–156.

Wei, Y. D., & Kim, S. (2002). widening inter-county inequality in Jiangsu Province, China, 1950–1995. *Journal of Development Studies, 38*(6), 142–164.

Cross-Sectional Studies on Mathematical Aptitude and Intelligent Quotient in North Eastern Tribes

Ratan Dasgupta

Abstract Cross-sectional studies are conducted on mathematical aptitude on a sample of 590 individuals from North Eastern tribes in Tripura over a time span of more than 5 years, viz., September 20, 2011–January 30, 2017. Scores on Intelligent Quotient (IQ) are obtained from interviews on 383 individuals over the years 2011–2014. The study is supplementary to the longitudinal study of Dasgupta (Longitudinal studies on mathematical aptitude and intelligence quotient in North Eastern Tribes, 2018). Total score, i.e., mathematical aptitude score plus IQ scores are also obtained on 383 individuals over the same time period of years 2011–2014, when traits mathematical aptitude score and IQ were assessed simultaneously. Scores are expressed in percentage when analyzed. Analyzed data indicate that mathematical aptitude score, intelligent quotient score, and sum-total of these two fluctuate over time, starting from high values after interaction with interviewer, when individuals are benefited from discussions in such assessment. The scores subsequently stabilize over time at a level slightly below the peak scores when the effect of interaction fades up. The average level of mathematical aptitude is low; although level of intelligence quotient score is comparatively high. Cross-sectional studies based on average scores at different time points with nonparametric regression indicate an upward trend of growth curve for all the traits, viz., IQ score, mathematical aptitude score, and total score, implying overall improvement of status over time. Income have a positive effect on education status. Individuals belonging to very-high-income group show high score in mathematical aptitude. Low mathematical aptitude score in early days of interview for individuals having good income is a possible indication of inherited or joint source of income, as observed in the study.

Keywords Mathematical aptitude · IQ · Proliferation rate · Cross-sectional study

R. Dasgupta (✉)
Theoretical Statistics and Mathematics Unit, Indian Statistical Institute,
203 B.T. Road, Kolkata 700108, India
e-mail: rdgupta@isical.ac.in; ratandasgupta@gmail.com

R. Dasgupta (ed.), *Advances in Growth Curve and Structural Equation Modeling*,
https://doi.org/10.1007/978-981-13-0980-9_5

MS Subject Classification: 62P15 · 62P25 · 62-07

1 Introduction and Methodology

Studies on mathematical aptitude and intelligence quotient in different ethnic groups are of interest. These may be conducted via longitudinal and cross-sectional studies. Whereas longitudinal studies are more informative, the cross-sectional studies are less expensive and less time consuming. As a result, the number of individuals that may be accommodated in cross-sectional studies are usually large compared to that for longitudinal studies. Some of the tribal individuals interviewed had higher education, each having their narration of success. Tribal individuals of other countries like American Indians and Alaska natives have similar pursuit toward higher learning, although apparently lagging behind the national norm, see e.g., Pavel and Inglebret (2007).

The number of tribal individuals interviewed on mathematical aptitude is 590 over the years 2011–17. In total, 383 individuals were interviewed on IQ questions

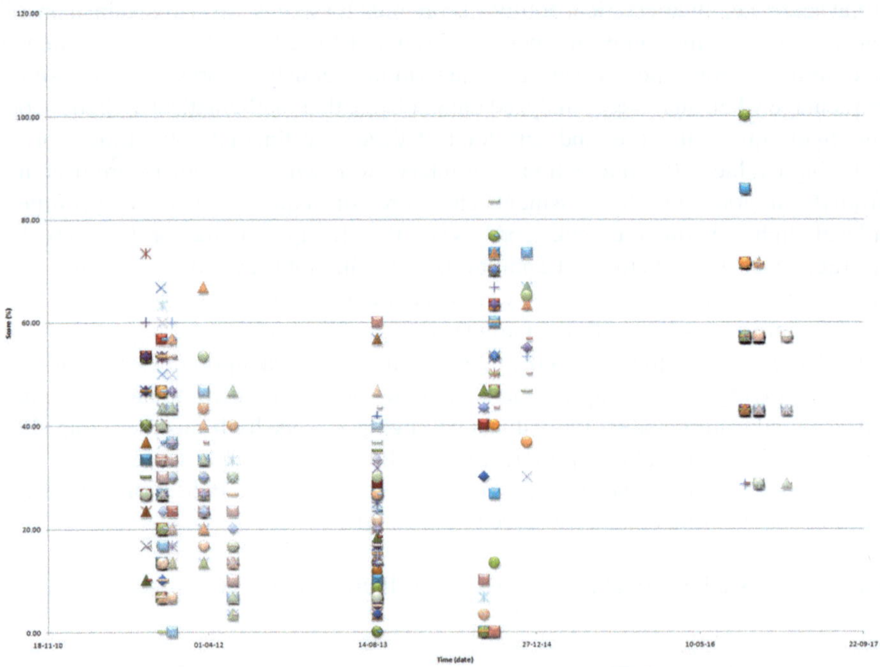

Fig. 1 Cross-sectional data (Math) of 590 tribal individuals. Scatter diagram of 590 tribal individuals' mathematical aptitude score in a time period of 6 years reveals an upward trend of mathematics aptitude score over time in Fig. 1

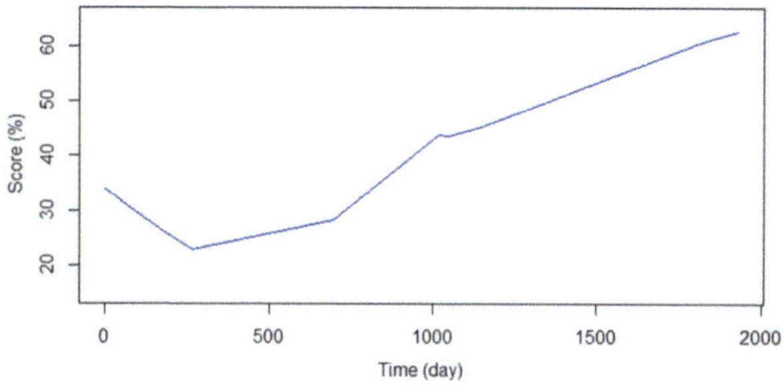

Fig. 2 Growth curve of cross-sectional data (Math) of 590 individuals. For each time point t when the interview is taken, we consider the median $m(t)$ of the mathematical aptitude scores recorded at that time. A LOWESS growth curve over these 18 pairs $(t, m(t))$ of time and median scores with $f = 2/3$ provides the growth curve is shown in Fig. 2. The curve has a downward trend at the initial stage. After that the curve shows an increasing upward trend over time. There is a cusp in the curve around 1000 days, and after this the rate of increase is a bit low

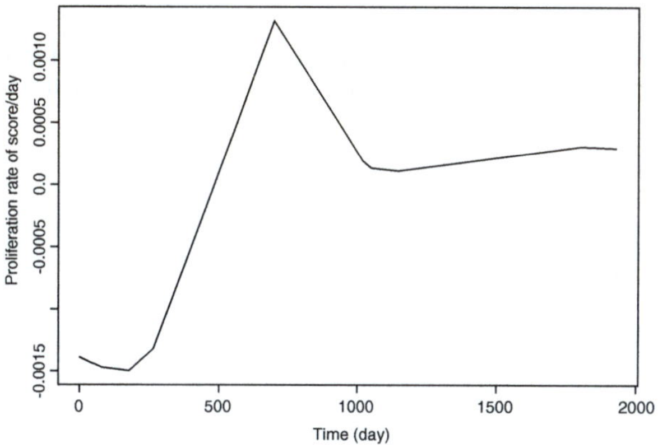

Fig. 3 Proliferation rate of 590 persons math score: wt. exp(−x); spline. Proliferation rate $\frac{d}{dt} \log y(t)$ over 18 time points for mathematical aptitude score y is computed based on a technique proposed in Dasgupta (2015), with exponentially decaying weights attached to empirical slopes computed from data pairs at different time points with respect to a fixed time point t of interest. More weights are given to data points near the time t of derivative computation, and less weights to distant time points from t. Weighted mean of the empirical slopes is considered at derivative stage of computation and the package smooth.spline with spar = 0.0001 at smoothing stage in SPlus provide the proliferation rate at the time point t. The modified procedure is explained in Fig. 10 of Dasgupta (2018). Consideration of median or trimmed mean of raw proliferation rates after exponential weighting as pooled estimate sometimes produces too smooth rates. The problem is circumvented by taking weighted averages as pooled estimate. A downward trend in the beginning is seen, and then a typical sharp rise over time is observed. Finally, stability of proliferation rate is seen toward the far end at a level below the peak and is shown in Fig. 3

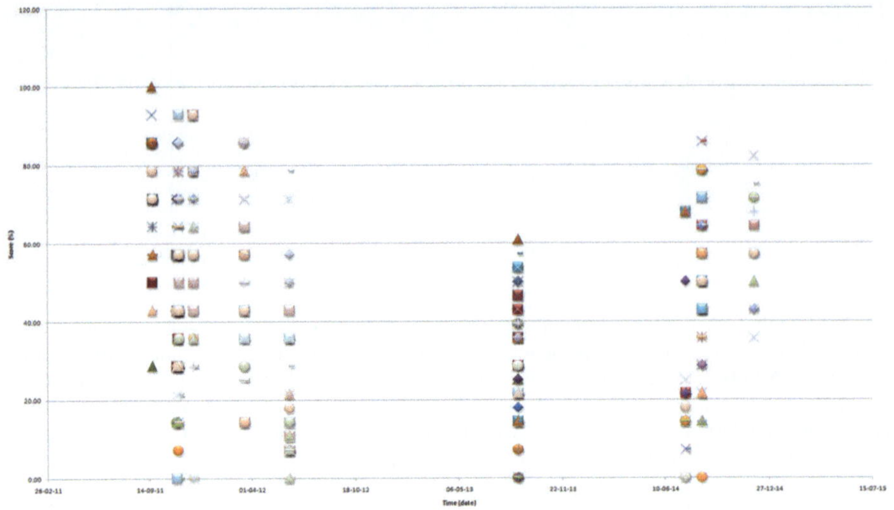

Fig. 4 Cross-sectional data (IQ) of 383 tribal individuals. Scatter diagram of 383 tribal individuals IQ score in a time period of 3 years reveals an initial downward trend, and then an upward trend is seen over time in Fig. 4

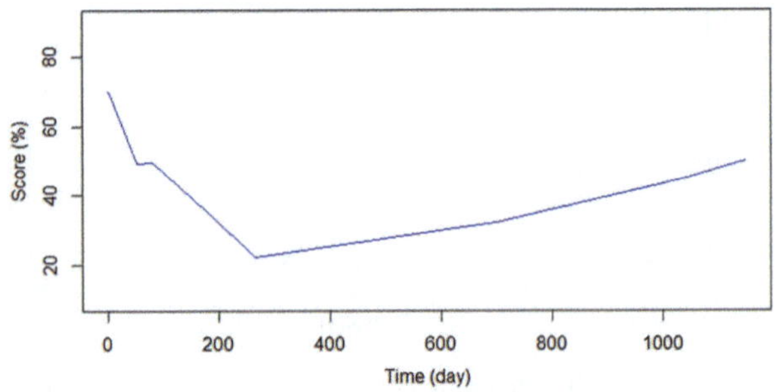

Fig. 5 Growth curve of cross-sectional data (IQ) of 383 individuals. For each time point where the interview is taken, we consider the median of the IQ scores recorded at that time. A LOWESS growth curve over these 14 pairs of values of time and median score with $f = 2/3$ provides the growth curve is shown in Fig. 5. The curve has an initial start from high level of score 70 due to advantage of interaction with the interviewer, then the curve shows a downward trend. After these initial stages, the curve consistently shows an increasing upward trend over time, reaching slowly at the score level 50 at the end

plus mathematical aptitude questions over the years 2011–14 in different regions in Tripura. Like the assessment procedure followed in longitudinal studies reported in Dasgupta (2018), the individuals in the cross-sectional studies are interviewed at different places of congress like students in different schools, individuals in their respective places of work, marketplaces, churches, and villages during data collection in the interiors of Tripura, following simple random sampling in a selection technique involving stratification.

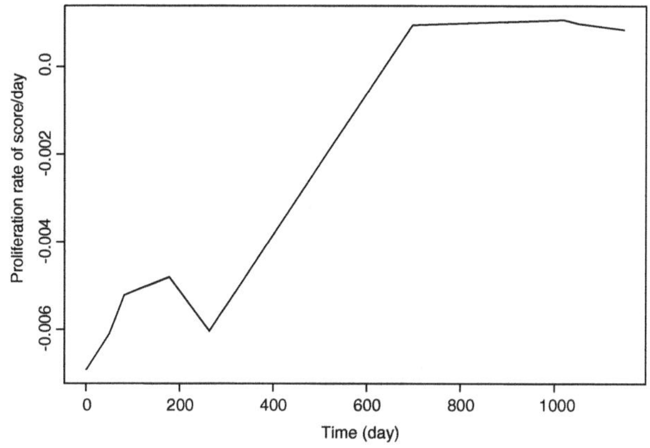

Fig. 6 Proliferation rate of 383 persons IQ score: wt. exp(−x); spline. Proliferation rate $\frac{d}{dt}\log y$ over 14 time points for IQ score y is computed based on a technique proposed in Dasgupta (2015), with exponentially decaying weights attached to empirical slopes computed from data pairs at different time points with respect to a fixed time point t of interest. More weights are given to data points near the time t of derivative computation, and less weights to distant time points from t. Weighted mean of the empirical slopes is considered at derivative stage of computation and the package smooth.spline with spar = 0.0001 at smoothing stage in SPlus provide the proliferation rate at the time point t. An upward trend in the beginning of the curve is seen, and then a sharp drop over time is observed; next a steep rise is seen. Finally, the curve stabilizes near the peak value toward the end and is shown in Fig. 6

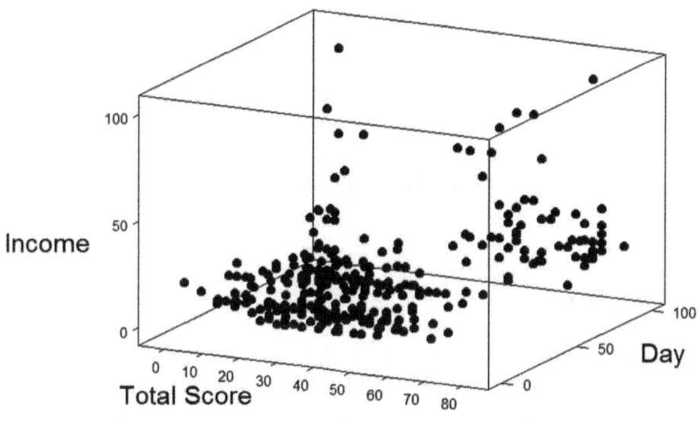

Fig. 7 Income versus total score and time. A three-dimensional picture on the variation of income, total score, and time is shown in Fig. 7, after scaling the maximum income (Rs. 25,000 per month) as 100. Day number 1148 is scaled to 100, as the maximum time duration. The individuals under longitudinal studies considered in Dasgupta (2018) are not included in this cross-sectional study. Income varies strongly with time and to a great extent with total score. Most tribal individuals have income below Rs. 12,500 per month, marked as 50 in the z axis. A few individuals with low total score have high income status; this status is possibly gathered from inheritance

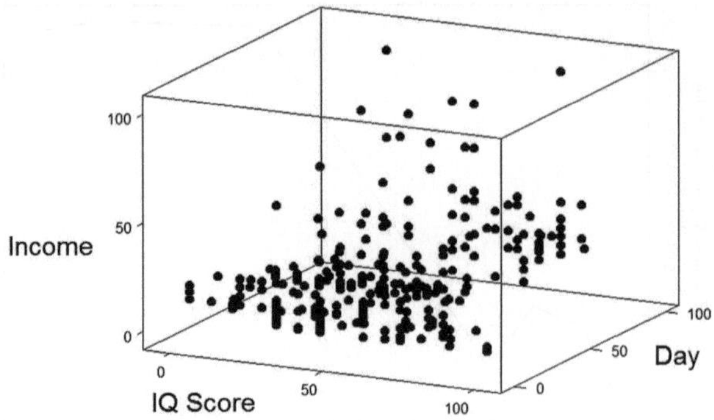

Fig. 8 Income versus IQ score and time. Here, the scaling of the variables income and time is as before. Three-dimensional picture on the variation of income, IQ score, and time, after scaling the maximum income indicates that the individuals with IQ score above 50 has a significant proportion with income more than Rs. 12,500 per month, i.e., above 50 mark in z axis. Time is a dominating factor influencing income to a great extent and IQ to some extent. For each fixed profile of IQ, there is a general tendency of an upturn in income over time. Very-high value of IQ indicates income above a minimum level, irrespective of time. The individuals under longitudinal study are not included and is shown in Fig. 8

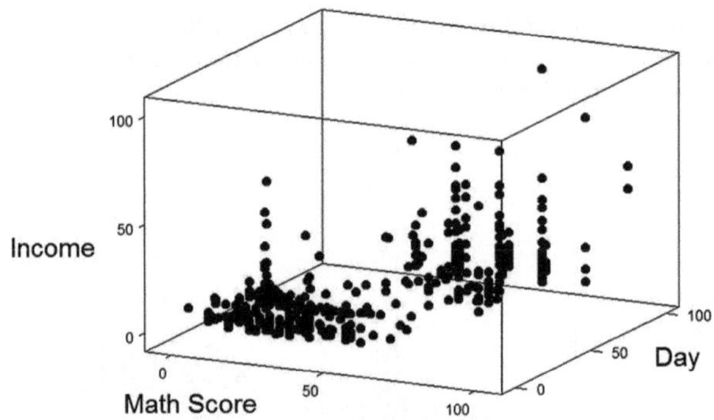

Fig. 9 Income versus math score and time. In Fig. 9, day number 1930 is scaled to 100. Apparently, there are two groups in the scatter. Individuals with mathematical aptitude score more than 50, in the scaled time region more than 50, i.e., above $1148/2 = 574$ days have a high percentage of well-to-do individuals. There are some individuals with low mathematical aptitude score in early days of interview having good income, possibly these individuals have inherited income components. The individuals under longitudinal study are not included

In Sect. 2, we state the results obtained from the data analyzed. Nonparametric LOWESS regression and spline smoothing are used to obtain the regression curves. Proliferation rates are obtained by a technique elaborated in Dasgupta (2018). The results are further discussed in Sect. 3.

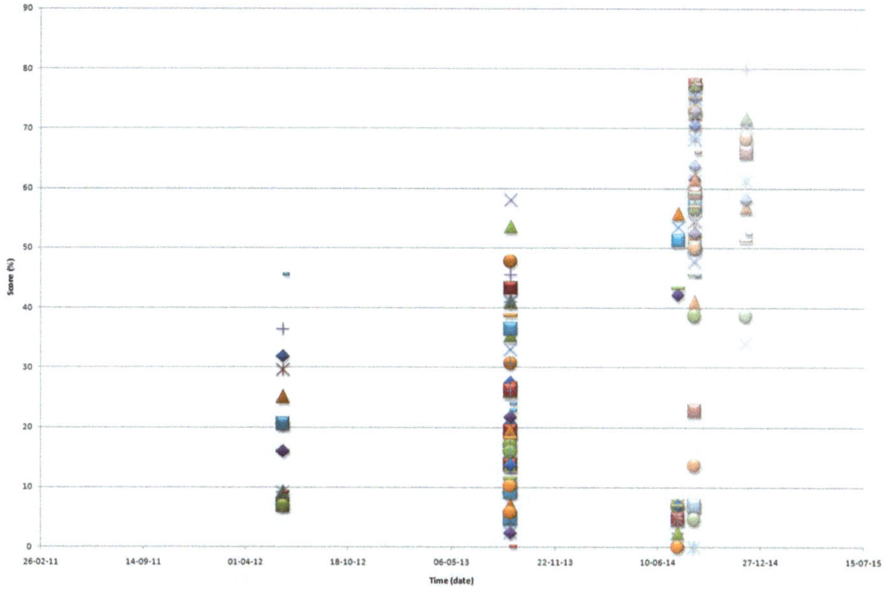

Fig. 10 Cross-sectional data (IQ+Math) of 383 individuals. Scatter diagram of 383 tribal individuals' total score in a time period of 3 years reveals an upward trend of IQ plus mathematical aptitude score over time in Fig. 10

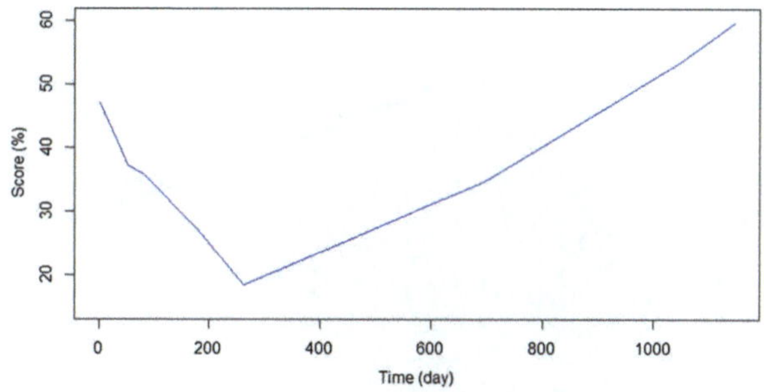

Fig. 11 Growth curve of cross-sectional data (IQ+Math) of 383 individuals. The curve shows a downward trend initially, after the effect of interaction with the interviewer fades out. The curve in Fig. 11 then shows a consistently sharp rise to reach the level of 60% total score

2 Data Analysis and the Results

The cross-sectional survey conducted over a time span of about 3 years revealed that IQ test score in percentage is at a slightly higher level than the mathematical aptitude score percentage on the same timescale. This feature is also revealed in the longitudinal analysis of the variables with less number of data.

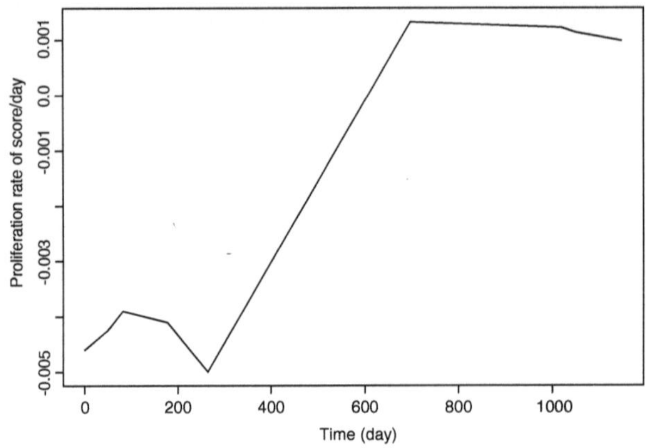

Fig. 12 Proliferation rate of 383 persons total score; wt. exp(−x); spline. Proliferation rate $\frac{d}{dt}\log y$ over 14 time points for total score y is computed based on a technique proposed in Dasgupta (2015), with exponentially decaying weights attached to empirical slopes computed from data pairs at different time points with respect to a fixed time point t of interest. More weights are given to data points near the time t of derivative computation, and exponentially decaying negligible weights are assigned to distant time points from t. Weighted mean of the empirical slopes is considered at derivative stage of computation and the package smooth.spline with spar = 0.0001 at smoothing stage in SPlus provide the proliferation rate at the time point t. The proliferation rate shows an initial uprise, and then it decreases to a lower level. A sharp uprise is seen thereafter, and that stabilizes toward far end with progress in time in Fig. 12. The curve is similar to shown in Fig. 6.

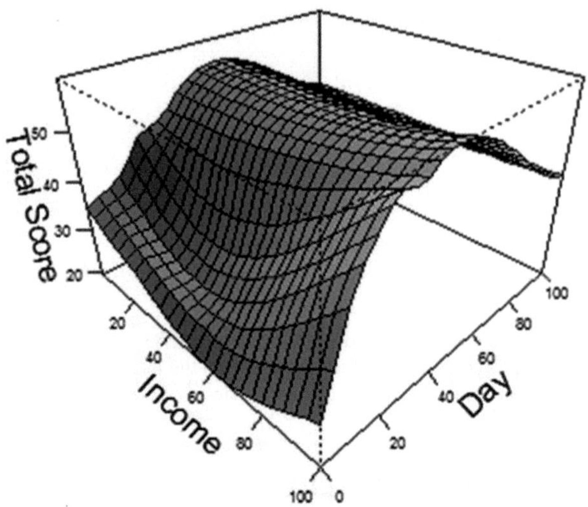

Fig. 13 Total score versus income and time. In Fig. 13, total score is seen as a function of income and time in a three-dimensional plot. For persons in lower income group with monthly income in the range [20, 60], expressed in percentage; the growth of total scores is slow for initial days within the range [0, 20]. Total score attains a high level around the time 75, after which a gradual decrease is seen. Recall that the variables are scaled to accommodate these in the finite range of graph

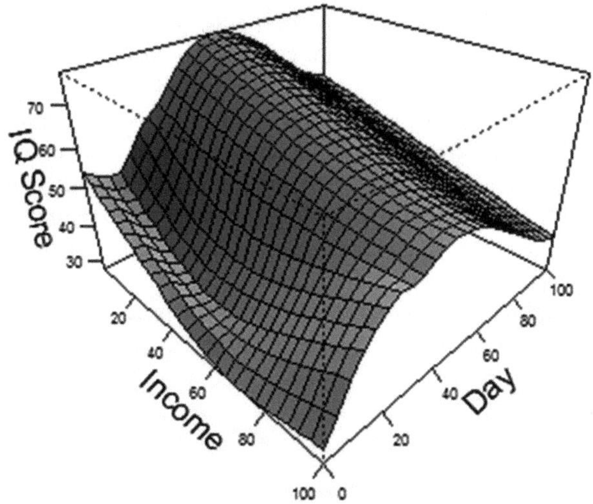

Fig. 14 IQ score versus income and time. In Fig. 14, variation of IQ score is shown as a function of income and time. In lower income group, the scores show downward trend for an initial period. After that the scores show an increasing trend over time. In this figure, the movement of scores over time toward the peak level reaching at about 80 days is relatively homogeneous for different income groups. This feature is absent in figure

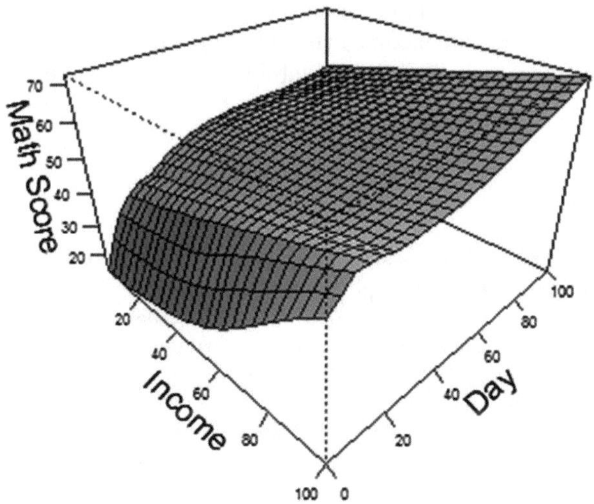

Fig. 15 Math score versus income and time. Figure 15 shows that for all income groups, the uprise in mathematical aptitude score is sharp from the beginning; due to individuals' interaction with the interviewer. Growth of scores then becomes moderate in the remaining time range exhibiting increasing trend till the far end. Individuals in very-high-income group over time then attain high mathematical aptitude score

Mathematical aptitude interviews in cross-sectional setup are conducted during the years 2011–17, whereas IQ tests are conducted during 2011–14. The IQ score and mathematics score are seen to be at the level of 50% and 45%, respectively, in LOWESS regression curve at the year 2014. The rate of increase in mathematical aptitude score is slightly lower toward the end. Level of education is low in lower income group. Individuals in very-high-income group show high mathematical aptitude score. Individuals with mathematical aptitude score more than 50, in the scaled time region more than 50, i.e., above 574 days have a high percentage of well-to-do individuals. Different features of collected data and growth curve-related results are explained in the figures with legends elaborating the procedure adopted.

3 Discussion

The number of individuals interviewed in the cross-sectional study is larger than that considered in Dasgupta (2018) for longitudinal study. There appears to be an association of mathematical aptitude with income among the persons interviewed in the tribal group. In general, there is an increasing trend in scores over time, suggesting the overall improvement of tribal lifestyle status. The level of IQ score is slightly above the mathematical aptitude score, as also seen in longitudinal studies. Total score and income are highly associated, thus there is an indication of high association of education with income in the tribal individuals interviewed. The curves on proliferation rates, in general, reach stability toward the end over time.

References

Dasgupta, R. (2015). Rates of convergence in CLT for two sample U-statistics in non iid case and multiphasic growth curve. In R. Dasgupta (Ed.), *Growth curve and structural equation modeling, Springer Proceedings in Mathematics & Statistics* (Vol. 132, pp. 35–58).

Dasgupta R. (2018). *Longitudinal studies on mathematical aptitude and intelligence quotient in North Eastern Tribes*. Appearing in this volume as chapter 1.

Pavel, M. D., & Inglebret, E. (2007). *The American Indian and Alaska Native Student's Guide to College Success*. London: Greenwood Publishing Group.

Growth Models for Repeated Measurement Mixture Experiments: Optimal Designs for Parameter Estimation and Growth Prediction

Manisha Pal, Nripes K. Mandal and Bikas K. Sinha

Abstract The present study focuses on the problems of parameter estimation and growth prediction in a quadratic growth model based on repeated measurements of growth, where the parameters in the model are assumed to be functions of 'treatments' which are treated as mixtures. The study concentrates not only on the optimality aspects of designs for most efficient estimation of the parameters but also on optimal prediction of growth at designated time points.

Keywords D- and A-optimal designs · Mixtures · Parameter estimation · Prediction · Quadratic growth model · Repeated measurements

1 Introduction

In an authoritative article, Sen (2013) has described in details some statistical perspectives of growth curve models (GCMs). As he has rightly narrated, Box (1950) initiated the study of growth (and wear) curves in simple biometric setups. Rao (1958, 1965) made significant contributions to GCM, while Potthoff and Roy (1964) systematically integrated GCMs in the mainstream of multivariate analysis of variance (MANOVA) and linked it to multivariate analysis of covariance (MANOCOVA). Von Rosen (1991) has provided a survey of papers which have influenced or have been influenced by the study of GCM due to Potthoff and Roy (1964).

Growth curve models are extensively used in economics, population studies, agriculture, biology, medical research, epidemiology, etc. to investigate growth over time. These models express growth as a function of time and repeated measurements of growth at discrete time points on the interval of study help to estimate the models. The existing literature on repeated measures experiments/designs covers a wide range of

M. Pal · N. K. Mandal
Calcutta University, Kolkata, India

B. K. Sinha (✉)
Indian Statistical Institute, Kolkata, India
e-mail: bikashsinha1946@gmail.com

R. Dasgupta (ed.), *Advances in Growth Curve and Structural Equation Modeling*,
https://doi.org/10.1007/978-981-13-0980-9_6

topics, starting from model formulations to data analyses to optimality studies. Shah and Sinha (1989) describe the basic results in this direction. See also Abt et al. (1997, 1998), Afsarinejad (1990).

In agriculture, crop growth models can be useful tools for optimizing farm management and integrating knowledge of crop physiology and climatic management. They provide information to farmers and policymakers on how to accomplish sustainable agriculture in the face of changing environment. Prediction of growth over time is also important for maximizing cropping system efficiency.

The growth of a crop depends on a number of factors, some of which are genetic and some environmental. While the farmer or producer has control over the genetic factor by his choice of the variety, the environmental factors, like temperature, rainfall, soil structure, soil quality, radiant energy, etc., are beyond his control. However, the productivity of the soil can be improved with the help of manure and fertilizers. These are basically mixtures of a number of ingredients, such as organic matter and nutrients like nitrogen, phosphorous, etc., the proportions of which in the mixture are very important to make the soil fertile. In a more general sense, the 'treatment' is being looked upon as a mixture.

In this paper, we undertake a study of the standard repeated measures models with a quadratic growth in time. Our main emphasis is the representation of so-called treatment effects as effects of mixtures of suitable 'components or ingredients' as in 'mixture experiments'. In so doing, we deal with models with correlated errors. We investigate the problems of finding D- and A-optimal designs for estimating the parameters of the model. We also deal with the prediction problem in such quadratic growth models.

2 Growth Model and Parameter Estimation

In its most general form, a growth curve model depicts the growth scenario of one or more experimental units [of the same nature], which is observed across the time domain, usually at equi-spaced time points within a given period of time. For the i-th experimental unit, the growth pattern is studied through its observed manifestation in the form of output vectors $Y_i = (y_{i1}, y_{i2}, \ldots, y_{in(i)})$ of dimension $n(i) \times 1$ at the distinct time points $t_{i1}; t_{i2}, \ldots, t_{in(i)}$ in that order. Naturally, the mean model of growth will depend on the actual nature of growth, usually assumed to be a polynomial in the time point t. Further, the degree of the polynomial is assumed to be the same across all plants under the purview of study.

For a p-th degree polynomial growth curve mean model, we may write

$$E(y_{ij}) = \beta_{0i} + t_{ij}\beta_{1i} + t_{ij}^2\beta_{2i} + \cdots + t_{ij}^p\beta_{pi}$$

for $i = 1, 2, \ldots, N; j = 1, 2, \ldots, n(i)$.

In the above, it is tacitly assumed that there are N experimental units and that the growth parameters, i.e. β-coefficients in the polynomial regression may depend on the units.

Further to this, the variance–covariance matrix of $\boldsymbol{Y}_i$ may be denoted by $\sigma^2\boldsymbol{W}(i)$, where $\boldsymbol{W}(i)$ is a non-diagonal positive definite correlation matrix. Various correlation structures may be imposed. For $n(i) \geq (p+1)$, estimability of the β-parameters is ensured irrespective of the nature of the $\boldsymbol{W}$-matrices. Data analysis and further studies are heavily dependent on the structures of the correlation matrices.

In this article, we confine to quadratic growth models, i.e. to $p=2$. Further to this, the 'treatments' applied to the experimental units are assumed to have been generated through mixtures in the sense of 'mixture experiments'. Thus, we contemplate a situation wherein we have q-component mixtures of the form $(x_1, x_2, \ldots, x_q)$ satisfying the conventional requirement: $0 \leq x_i \leq 1,\ i = 1, 2, \ldots, q, \sum_{i=1}^{q} x_i = 1$. In the context of an agricultural experiment, therefore, the i-th plant [singly as an entity or in the form of a 'bush' labelled i] may receive a 'treatment mixture' which may be identified as $(x_{i1}, x_{i2}, \ldots, x_{iq})$, for which $0 \leq x_{ij} \leq 1,\ j = 1, 2, \ldots, q, \sum_{j=1}^{q} x_{ij} = 1$. Consequently, the underlying β-coefficients in a quadratic regression model may be represented as linear or quadratic functions of $(x_{i1}, x_{i2}, \ldots, x_{iq})$ as given below:
Linear:

$$\begin{aligned}
\beta_{0i} &= \alpha_{01}x_{i1} + \alpha_{02}x_{i2} + \cdots + \alpha_{0q}x_{iq};\\
\beta_{1i} &= \alpha_{11}x_{i1} + \alpha_{12}x_{i2} + \cdots + \alpha_{1q}x_{iq};\\
b_{2i} &= \alpha_{21}x_{i1} + \alpha_{22}x_{i2} + \cdots + \alpha_{2q}x_{iq}.
\end{aligned}$$

Quadratic:

$$\begin{aligned}
\beta_{0i} &= \alpha_{01}x_{i1} + \alpha_{02}x_{i2} + \cdots + \alpha_{0q}x_{iq} + \alpha_{012}x_{i1}x_{i2} + \cdots + \alpha_{01q}x_{i1}x_{iq} + \cdots + \alpha_{0q-1,q}x_{i,q-1}x_{iq};\\
\beta_{1i} &= \alpha_{11}x_{i1} + \alpha_{12}x_{i2} + \cdots + \alpha_{1q}x_{iq} + \alpha_{112}x_{i1}x_{i2} + \cdots + \alpha_{11q}x_{i1}x_{iq} + \cdots + \alpha_{1,q-1,q}x_{i,q-1}x_{iq};\\
\beta_{2i} &= \alpha_{21}x_{i1} + \alpha_{22}x_{i2} + \cdots + \alpha_{2q}x_{iq} + \alpha_{212}x_{i1}x_{i2} + \cdots + \alpha_{21q}x_{i1}x_{iq} + \cdots + \alpha_{2,q-1,q}x_{i,q-1}x_{iq}.
\end{aligned}$$

Our interest lies in the estimation of the coefficients α's in the ultimate mixture model for a quadratic response growth curve. There are a total of $3q$ (for linear) or $3 \times {}^{q+1}C_2$ (for quadratic) parameters, apart from the error variance term σ^2. It must be remembered that for each experimental unit, the observations across the time domain are 'correlated' in some form. To achieve simplicity in our subsequent analysis and optimality study, we may assume the same structure of this correlation matrix for all units. In this study, we further confine to mixtures with three components, i.e. with $q=3$. Essentially, then, we are in the framework of a linear correlated model with 9 or 18 parameters, depending on the type of representation of the β-coefficients in terms of the mixing proportions.

We, therefore, have the following mean growth model given by

$$\eta(t) = \beta_0 + \beta_1 t + \beta_2 t^2,$$

where, by suitable transformation, the range of t can be taken to be $[-1, 1]$.

The parameters β_0, β_1, β_2 are (i) linear or (ii) quadratic functions of the mixture combination, namely,

$$\beta_i \equiv \beta_i(\boldsymbol{x}) = \sum_{i=1}^{q} \alpha_{ij} x_j, \quad i = 0, 1, 2, \text{ in (i)}$$

and

$$\beta_i \equiv \beta_i(\boldsymbol{x}) = \sum_{j=1}^{q} \alpha_{ij} x_j + \sum_{j<k=1}^{q} \alpha_{ijk} x_j x_k, \quad i = 0, 1, 2, \text{ in (ii)},$$

where $\boldsymbol{x} = (x_1, x_2, \ldots, x_q)$.

That is,

$$\beta_i(\boldsymbol{x}) = \mathrm{g}'(\boldsymbol{x})\boldsymbol{\alpha}^{(i)}, \quad i = 0, 1, 2,$$

where

$$\begin{aligned} \mathbf{g}(\boldsymbol{x}) &= (x_1, x_2, \ldots, x_q)', \text{ in (i)} \\ &= (x_1, x_2, \ldots, x_q, x_1x_2, \ldots, x_1x_q, \ldots, x_{q-1}x_q)', \text{ in (ii)} \end{aligned}$$

and

$$\begin{aligned} \boldsymbol{\alpha}^{(i)} &= (\alpha_{i1}, \alpha_{i2}, \ldots, \alpha_{iq})', \text{ in (i)} \\ &= (\alpha_{i1}, \alpha_{i2}, \ldots, \alpha_{iq}, \alpha_{i12}, \ldots, \alpha_{i1q}, \alpha_{i23}, \ldots, \alpha_{2q}, \ldots, \alpha_{iq-1,q})', \text{ in (ii).} \end{aligned}$$

Hence, the full mean growth model is given by

$$\begin{aligned} \eta(t; \boldsymbol{x}) &= \boldsymbol{g}(\boldsymbol{x})'\boldsymbol{\alpha}^{(0)} + \boldsymbol{g}(\boldsymbol{x})'\boldsymbol{\alpha}^{(1)}t + \boldsymbol{g}(\boldsymbol{x})'\boldsymbol{\alpha}^{(2)}t^2 \\ &= \left[(1, t, t^2) \otimes \boldsymbol{g}(\boldsymbol{x})'\right]\left(\boldsymbol{\alpha}^{(0)'}, \boldsymbol{\alpha}^{(1)'}, \boldsymbol{\alpha}^{(2)'}\right)', \\ &= \boldsymbol{f}(t; \boldsymbol{x})'\boldsymbol{\theta}, \text{ say,} \end{aligned} \tag{1}$$

where $\boldsymbol{f}'(t; \boldsymbol{x}) = (1, t, t^2) \otimes g(\boldsymbol{x})'$, $\boldsymbol{\theta} = (\boldsymbol{\alpha}^{(0)'}, \boldsymbol{\alpha}^{(1)'}, \boldsymbol{\alpha}^{(2)'})'$.

Our problem is now to find an optimal design for estimating $\boldsymbol{\theta}$ based on repeated observations on growth.

3 Mixture Models

Let $\boldsymbol{x} = (x_1, x_2, \ldots, x_q)'$ denote the vector of proportions of q mixing components and $\varsigma(\boldsymbol{x})$ be the corresponding mean response. The factor space is a simplex, which is given by

$$\Xi = \left\{ (x_1, x_2, \ldots, x_q) \middle| x_i \geq 0, i = 1, 2, \ldots, q; \sum_{i=1}^{q} x_i = 1 \right\}. \tag{2}$$

Scheffé (1958) introduced the following models in canonical forms of different degrees to represent the mean response function $\varsigma(\boldsymbol{x})$:

$$\text{Linear}: \quad \varsigma(\boldsymbol{x}) = \sum_{i=1}^{q} \beta_i x_i \tag{3}$$

$$\text{Quadratic}: \quad \varsigma(\boldsymbol{x}) = \sum_{i=1}^{q} \beta_i x_i + \sum_{i<j=1}^{q} \beta_{ij} x_i x_j \tag{4}$$

$$\text{Cubic}: \quad \varsigma(\boldsymbol{x}) = \sum_{i=1}^{q} \beta_i x_i + \sum_{i<j=1}^{q} \beta_{ij} x_i x_j + \sum_{i<j<k=1}^{q} \beta_{ijk} x_i x_j x_k \tag{5}$$

$$\text{Special cubic}: \quad \varsigma(\boldsymbol{x}) = \sum_{i=1}^{q} \beta_i x_i + \sum_{i<j=1}^{q} \beta_{ij} x_i x_j + \sum_{i<j<k=1}^{q} \beta_{ijk} x_i x_j x_k + \sum_{i<j=1}^{q} \delta_{ij} x_i x_j (x_i - x_j) \tag{6}$$

In the above, generic notations have been used for the model parameters in different versions of mixture models.

4 Optimum Designs

In our study for the estimation of the model parameters of (2) for any given t, we consider an approximate or continuous mixture design given by

$$\xi = \left\{ \begin{matrix} \boldsymbol{x}_{(1)} & \boldsymbol{x}_{(2)} & \ldots & \boldsymbol{x}_{(N)} \\ w_1 & w_2 & \ldots & w_N \end{matrix} \right\}, \tag{7}$$

where $\boldsymbol{x}_{(1)}, \boldsymbol{x}_{(2)}, \ldots, \boldsymbol{x}_{(N)}$ are the distinct support points of the design ξ with masses w_1, w_2, ..., w_N, respectively, such that $w_i \geq 0$, $j = 1, 2, \ldots, N$, and $\sum_{i=1}^{N} w_i = 1$. The information matrix of the estimate $\hat{\boldsymbol{\theta}}$ of $\boldsymbol{\theta}$ for the design ξ is

$\boldsymbol{M}(\xi) = \sum_{i=1}^{N} w_i \, \boldsymbol{f}(t; \boldsymbol{x}_{(i)}) \, \boldsymbol{f}(t; \boldsymbol{x}_{(i)})'$. We restrict to the class of designs for which *the information matrix* $\boldsymbol{M}(.)$ is positive definite, so that all the parameters of the model (2) are estimable.

Suppose the growth of the crop is observed at k time points $t_1 < t_2 < \ldots < t_k$ in the range $[-1, 1]$. Then, a typical experimental design d in our study will involve an approximate mixture design ξ along with k time points $t_1, t_2, \ldots, t_k$ at which observations are to be taken. Let us denote by $\boldsymbol{D}_k$ the class of all such designs. Then, our problem is to find an optimal design in $\boldsymbol{D}_k$.

It is easy to note that the time points remain invariant w.r.t. the choice of the mixing proportions in the treatment. Hence, we can use the de la Garza phenomenon given below to reduce the number of time points at which growth is to be observed:

Theorem 4.1 *(de la Garza Phenomenon) Under a $(p-1)$th degree polynomial regression model, any optimal design can be based on at most p design points, the minimum number of support points such that all parameters are estimable (cf. Garza 1954).*

By virtue of the above Theorem, for a quadratic (in t) growth model, it is enough to concentrate on ***three*** distinct time points, namely $t_1 < t_2 < t_3$, in the range $[-1, 1]$.

Optimality, therefore, refers to the optimum choices of $t_1 < t_2 < t_3$ and the mixture design ξ.

Hence, our search for the optimal design will be made in the class $\boldsymbol{D}_3$.

Let $\boldsymbol{Y} = (Y_1, Y_2, Y_3)'$ be the vector of observed growths of the crop at t_1, t_2, t_3, and let the dispersion matrix of $\boldsymbol{Y}$ be $\boldsymbol{\Sigma}$. Let us assume that

$$\boldsymbol{\Sigma} = \sigma^2 \begin{bmatrix} 1 & \rho^{(12)} & \rho^{(13)} \\ \rho^{(12)} & 1 & \rho^{(23)} \\ \rho^{(13)} & \rho^{(23)} & 1 \end{bmatrix}, \tag{8}$$

where the common error variance is σ^2 and $\rho^{(ij)} = \text{corr}(Y_i, Y_j), \quad i, j = 1(1)3, \quad i < j$.

Then, the information matrix of any design $d \in \boldsymbol{D}_3$ is given by

$$\begin{aligned} \boldsymbol{M}(d) &= \sum_{i=1}^{n} w_i \left[\boldsymbol{f}(t_1, \boldsymbol{x}^{(i)}), \boldsymbol{f}(t_2, \boldsymbol{x}^{(i)}), \boldsymbol{f}(t_3, \boldsymbol{x}^{(i)}) \right] \boldsymbol{\Sigma}^{-1} \begin{bmatrix} \boldsymbol{f}'(t_1, \boldsymbol{x}^{(i)}) \\ \boldsymbol{f}'(t_2, \boldsymbol{x}^{(i)}) \\ \boldsymbol{f}'(t_3, \boldsymbol{x}^{(i)}) \end{bmatrix} \\ &= \sum_{i=1}^{n} w_i \left\{ \begin{bmatrix} 1\,1\,1 \\ t_1\,t_2\,t_3 \\ t_1^2\,t_2^2\,t_3^2 \end{bmatrix} \otimes \boldsymbol{g}(\boldsymbol{x}^{(i)}) \right\} \boldsymbol{\Sigma}^{-1} \left\{ \begin{bmatrix} 1\,1\,1 \\ t_1\,t_2\,t_3 \\ t_1^2\,t_2^2\,t_3^2 \end{bmatrix} \otimes \boldsymbol{g}(\boldsymbol{x}^{(i)}) \right\}' \\ &= \boldsymbol{M}_1(\boldsymbol{t}) \otimes \boldsymbol{M}_2(\xi), \end{aligned} \tag{9}$$

where

$$M_1(t) = \begin{bmatrix} 1 & 1 & 1 \\ t_1 & t_2 & t_3 \\ t_1^2 & t_2^2 & t_3^2 \end{bmatrix} \Sigma^{-1} \begin{bmatrix} 1 & 1 & 1 \\ t_1 & t_2 & t_3 \\ t_1^2 & t_2^2 & t_3^2 \end{bmatrix}' = T\Sigma^{-1}T', \tag{10}$$

$$M_2(\xi) = \sum_{i=1}^{n} w_i g(x^{(i)}) g(x^{(i)})', \tag{11}$$

$$T = \begin{bmatrix} 1 & 1 & 1 \\ t_1 & t_2 & t_3 \\ t_1^2 & t_2^2 & t_3^2 \end{bmatrix}.$$

Clearly, $M_2(\xi)$ is the information matrix of a continuous mixture design ξ given by (8) for estimating the parameter vector γ of a mixture model

$$\varsigma(x) = g(x)'\gamma. \tag{12}$$

For estimating γ, a necessary and sufficient condition for a design to be D- or A-optimal is given by Kiefer's (1974) Equivalence Theorem, which also helps to get an idea of the design points. For D- and A-optimality designs, the Theorem is given as follows:

Theorem 3.2 *(Equivalence Theorem) A necessary and sufficient condition for a design ξ^* to be D-optimal (A-optimal) for estimating the parameters of the model (10) is that*

$$\begin{aligned} &g(x)'M^{-1}(\xi^*)g(x) \le p \\ &(g(x)'M^{-2}(\xi^*)g(x) \le Trace\{M^{-1}(\xi^*)\} \text{ for all } x \in \Xi, \end{aligned} \tag{13}$$

where p denotes the number of parameters to be estimated. Equality holds at the support points of ξ^*.

For a design ξ given by (3), consider

$$d_1(x) = g(x)'M^{-1}(\xi)g(x) - p$$

and $d_2(x) = g(x)'M^{-2}(\xi^*)g(x) - Trace\{M^{-1}(\xi^*)\}$.

For a linear mixture model, where $g(x) = (x_1, x_2, \ldots, x_q)'$, $d_1(x)$ is a quadratic convex function in x, M (ξ) is positive definite. Hence, it can have at most two maximal points with respect to each component of x, namely, at the two extremes x_i = 0, 1 for all i = 1, 2, …, q. Thus, because of the restriction $\sum_{i=1}^{q} x_i = 1$, the support points of the optimal design will be of the form (1, 0, …, 0) and its permutations, that is, $d_1(x)$ is maximized at the extremes points of Ξ. For quadratic mixture model, where $g(x) = (x_1, x_2, \ldots, x_q, x_1x_2, \ldots, x_1x_q, \ldots, x_{q-1}x_q)'$, $d_1(x)$ is the fourth-degree polynomial in x. Hence, it can have at most three maximal points with respect to each component of $\boldsymbol{x}$, namely, at the two extremes and one in-between. Since $d_1(\boldsymbol{x})$ is invariant with respect to the components of $\boldsymbol{x}$, the maximizing points will also be invariant with respect to the components of $\boldsymbol{x}$, that is, they will be at the barycentres of Ξ. Barycentres of depth j are defined as points in Ξ with $(j+1)$ non-zero components each having value $1/(j+1)$, and the remaining components are all zero, $0 \leq j \leq q-1$. Thus, the extreme points of Ξ are the barycentres of depth zero.

Arguing in the same way as before with $d_1(\boldsymbol{x})$ replaced by $d_2(\boldsymbol{x})$, we have that the support points of an A-optimal design will be at the extremes points of Ξ for a linear mixture model and will be at the barycentres of Ξ for a quadratic mixture model.

Having obtained the support points of the optimal designs, the associated masses can be obtained by maximizing det[$\boldsymbol{M}(\xi)$] in D-optimality criterion, and by minimizing $Trace\{\boldsymbol{M}^{-1}(\xi)\}$ in A-optimality criterion. The optimal designs are obtained as follows (cf. Galil and Kiefer 1977; Sinha et al. 2014):

(a) For linear mixture model, the D- and A-optimal designs are identical, each being a saturated design with support points at the extreme of the simplex Ξ each with mass $1/q$.

(b) For quadratic mixture model, the D-optimal design is a saturated design with support points at the extreme points and at (1/2, 1/2, 0, …, 0)' and its permutations, each with mass $1 / {}^{q+1}C_2$. For $q \neq 3$, the A-optimal design has the same support points, but the mass is π_0/q at each of the extreme points of Ξ, and $(1 - \pi_0) / {}^{q}C_2$ at each of the points (1/2, 1/2, 0, …, 0) and its permutations, where $\pi_0 = \frac{\sqrt{q(q+8C(q,2))}}{\sqrt{q(q+8C(q,2))+4C(q,2)}}$, while for $q=3$ the support points are the extreme points of Ξ each with mass 0.1417, the points and its permutations each with mass 0.1873, and the point (1/3, 1/3, 1/3)' with mass 0.0130.

4.1 D-Optimal Design

To obtain the D-optimal design for the present study, we note that for any design $d \in \boldsymbol{D}_3$, $\det(\boldsymbol{M}(d)) = \det(\boldsymbol{M}_1(\boldsymbol{t})) \times \det(\boldsymbol{M}_2(\xi))$, where $\boldsymbol{M}_1(\boldsymbol{t})$ and $\boldsymbol{M}_2(\xi)$ are given by (10) and (11), respectively. Thus, the D-optimal design will be a combination of the D-optimal design given by (a) above and (t_1, t_2, t_3) that maximizes $\det(\boldsymbol{M}_1(\boldsymbol{t}))$.

Result $\det(\boldsymbol{M}_1(\boldsymbol{t}))$ is maximized at $t_1 = -1, t_2 = 0, t_3 = 1$.

Proof From (10), $\det(\boldsymbol{M}_1(\boldsymbol{t})) = \det(\boldsymbol{T}\boldsymbol{\Sigma}^{-1}\boldsymbol{T}') = \det(\boldsymbol{\Sigma}^{-1}) \times \det(\boldsymbol{T}\boldsymbol{T}')$ which, after some algebra, reduces to

$$\det(\boldsymbol{M}_1(\boldsymbol{t})) = \det(.\boldsymbol{\Sigma}^{-1})[(t_3 - t_1)(t_3 - t_2)(t_2 - t_1)].$$

To maximize $[(t_3 - t_1)(t_3 - t_2)(t_2 - t_1)]$ by suitable choices of the t's such that $-1 \leq t_1 < t_2 < t_3 \leq 1$, we note that for any (fixed) choice of (t_2, t_3), t_1 must be the smallest, i.e. $t_1 = -1$. That simplifies $[(t_3 - t_1)(t_3 - t_2)(t_2 - t_1)$ to $[(t_3 + 1)(t_2 + 1)(t_3 - t_2)$.

Next for any (fixed) choice of t_2, t_3 must be the highest, i.e. $t_3 = +1$. That reduces $(t_3 - t_1)(t_3 - t_2)(t_2 - t_1)$ to $[2(1+t_2)\,(1-t_2)]$, whence $t_2 = 0$ is the best choice. Thus, $[-1, 0, +1]$ produces maximum value of det $\boldsymbol{M}_1(t)$. This is true irrespective of the nature of $\boldsymbol{\Sigma}$, the correlation structure of the errors.

4.2 A-Optimal Design

To obtain the A-optimal design for estimating the parameters of the model (2), we have from (9) that for any design $d \in \boldsymbol{D}_3$,

$$Trace\left[\boldsymbol{M}(d)^{-1}\right] = Trace\left[\boldsymbol{M}_1(t)^{-1}\right] Trace\left[\boldsymbol{M}_2(\xi)^{-1}\right].$$

Again, since $\boldsymbol{M}_1(\boldsymbol{t})$ involves $\boldsymbol{\Sigma}$ and (t_1, t_2, t_3), while $\boldsymbol{M}_2(\xi)$ is the information matrix of the mixture design ξ, the A-optimal design will be a combination of the A-optimal design for estimating the parameters of the mixture model (10), which is given in (b) above, and (t_1, t_2, t_3) that minimizes $Trace\left(\boldsymbol{M}_1(\boldsymbol{t})^{-1}\right)$.

When $\rho^{(ij)}$'s in Σ are all equal, we have the following result:

Result 4.2.1 When $\rho^{(ij)} = \rho$, for $1 \le i < j \le 3$, the optimal choice of t_1, t_2, t_3 is $t_1 = -1$, $t_3 = 1$, $t_2 = 0$, whatever be ρ.

Proof Noting that $\det(T) = (t_3 - t_1)(t_3 - t_2)(t_2 - t_1)$, we have

$$\begin{aligned} Trace\left(\boldsymbol{M}_1(t)^{-1}\right) &= Trace\left[\Sigma(\boldsymbol{T}'\boldsymbol{T})^{-1}\right. \\ &= \sigma^2\left\{1 + \frac{2(1-\rho^{(12)})[1 + (t_3+t_2)(t_3+t_1) + t_1 t_2 t_3^2]}{(t_3-t_1)(t_3-t_2)(t_2-t_1)^2}\right. \\ &\quad \left. - \frac{2(1-\rho^{(13)})[1 + (t_3+t_2)(t_2+t_1) + t_1 t_2^2 t_3]}{(t_3-t_1)^2(t_3-t_2)(t_2-t_1)} + \frac{2(1-\rho^{(23)})[1 + (t_3+t_1)(t_2+t_1) + t_1^2 t_2 t_3]}{(t_3-t_1)(t_3-t_2)^2(t_2-t_1)}\right\}. \end{aligned}$$

For $\rho^{(ij)} = \rho,\ 1 \le i < j \le 3$, the above expression reduces to

$$Trace\left(\boldsymbol{M}_1(t)^{-1}\right) = \sigma^2\left\{1 + 2(1-\rho)\left[\frac{1 + (t_3+t_2)^2 + t_1 t_2 t_3}{(t_3-t_1)^2(t_2-t_1)^2} + \frac{1 + (t_3+t_1)(t_2+t_1) + t_1^2 t_2 t_3}{(t_3-t_1)(t_3-t_2)^2(t_2-t_1)}\right]\right\}. \qquad (13)$$

It is easy to note that (13) is an increasing function of t_1. Hence, *Trace* $\left(\boldsymbol{M}_1(\boldsymbol{t})^{-1}\right)$ is minimized at $t_1 = -1$. Again, we can express (13) as

$$\sigma^2\left\{1 + 2(1-\rho)\left[\frac{1/t_3^2 + (+t_2/t_3)^2 + t_1 t_2/t_3}{(1 - t_1/t_3)^2(t_2-t_1)^2} + \frac{1 + (1 + t_1/t_3)(t_2+t_1) + t_1^2 t_2}{t_3(1 - t_1/t_3)(1 - t_2/t_3)^2(t_2-t_1)}\right]\right\},$$

which shows that $Trace\left(\boldsymbol{M}_1(\boldsymbol{t})^{-1}\right)$ is a decreasing function of t_3. Hence, $Trace\left(\boldsymbol{M}_1(\boldsymbol{t})^{-1}\right)$ is minimized at $t_3 = 1$.

Now for $t_1 = -1$, $t_3 = 1$, we get

Table 1 Optimum choice of (t_1, t_2, t_3) for some combinations of $(\rho^{(12)}, \rho^{(13)}, \rho^{(23)})$

$\rho^{(12)}$	$\rho^{(13)}$	$\rho^{(23)}$	Optimum		
			t_1	t_2	t_3
0.25	0.75	0.90	−1	0.3236	1
0.25	0.75	0.75	−1	0.1811	1
0.25	0.75	0.65	−1	0.1253	1
0.25	0.10	0.25	−1	−0.0112	1

$$Trace\big(\boldsymbol{M}_1(\boldsymbol{t})^{-1}\big) = \sigma^2\left\{1 + (1-\rho)\left[\frac{1}{(1+t_2)^2} + \frac{1}{(1-t_2)^2}\right]\right\},$$

which is a convex function of t_2 since $\frac{d^2}{dt_2^2}Trace\big(\boldsymbol{M}_1^{-1}(\xi)\big) = 3\sigma^2\left[\frac{1}{(1+t_2)^4} + \frac{1}{(1-t_2)^4}\right] > 0$.

Hence, the optimal value of t_2 is obtained by solving $\frac{d}{dt_2}Trace\big(\boldsymbol{M}_1^{-1}(\xi)\big) = \frac{\sigma^2(1-\rho)}{2}\left[\frac{1}{(1-t_2)^3} - \frac{1}{(1+t_2)^3}\right] = 0$, which gives $t_2 = 0$.

For $\rho^{(ij)}$'s not all equal, it is difficult to algebraically find the optimal choices of $t_1 < t_2 < t_3$. However, numerical calculation with several combinations of $\big(\rho^{(12)}, \rho^{(13)}, \rho^{(23)}\big)$ shows that $t_1 = -1, t_3 = 1$, whatever be $\big(\rho^{(12)}, \rho^{(13)}, \rho^{(23)}\big)$, but t_2 depends on the values of $\rho^{(ij)}$'s. Table 1 shows the optimum values of $t_1 < t_2 < t_3$ for some combinations of $\rho^{(ij)}$'s.

Remark It is well known that for a saturated model, the nature of D-optimal design is invariant with respect to the choice of the correlation matrix. On the other hand, for A-optimality criterion, the criterion function depends explicitly on the correlation parameters in a rather complicated form. For $q = 3$, we have exhibited the nature of optimal designs for specific choices of these parameters. Abt et al. (1997, 1998) discussed optimality issues for two types of correlation matrices, viz., the intra-class and AR(1) underlying a quadratic growth model. We did not venture into the study of A-optimality with respect to AR(1) model in this paper.

5 Optimum Designs for Prediction

In the rest of this article, we confine to mixtures with three components, i.e. $q = 3$, and we turn to the problem of prediction of the yield for a designated mixture of the three components, viz., $(a; b; c)$—assumed to have not been covered by the mixture experiment. Since unbiased prediction of 'yield' relates to unbiased estimation of the 'mean yield', it may be noted that for any $q \geq 3$, there is an infinite class of mixture designs providing unbiased estimates of the model parameters and hence of the mean yield. The simplest such mixture design for $q = 3$ is given by

$$\xi_0 = [(1;0;0)P;\ (0;1;0)Q;\ (0;0;1)R;\ 0 < P;\ Q;\ R < 1;\ P + Q + R = 1].$$

All the model parameters are trivially unbiasedly estimated based on the observations collected at each of the three time points, viz., $t = -1; 0; 1$. For an HLMM, it has been established that optimal estimation of all the nine model parameters under a mixture design rests on exclusive use of the above three mixtures, viz., (1; 0; 0); (0; 1; 0); (0; 0; 1) with suitable weights.

Denote by $\bar{y}_{(1,0,0)}(-1)$ the average 'yield' at the time point $t = -1$ based on $NP/3$ observations corresponding to the mixture (1; 0; 0), where N represents the total number of observations in the experiment and P is the mixing proportion attached to the mixture (1; 0; 0). Its model expectation is given by $\alpha_{01} - \alpha_{11} + \alpha_{21}$. Likewise, at the other two time points, $E\left[\bar{y}_{(1,0,0)}(0)\right] = \alpha_{01}$ and $E\left[\bar{y}_{(1,0,0)}(1)\right] = \alpha_{01} + \alpha_{11} + \alpha_{21}$. From these, we deduce that

$$\begin{aligned}
\hat{\alpha}_{01} &= \bar{y}_{(1,0,0)}(0);\\
2\hat{\alpha}_{21} &= \bar{y}_{(1,0,0)}(+1) + \bar{y}_{(1,0,0)}(-1) - 2\bar{y}_{(1,0,0)}(0)\\
2\hat{\alpha}_{11} &= \bar{y}_{(1,0,0)}(+1) - \bar{y}_{(1,0,0)}(-1).
\end{aligned}$$

Therefore, under the equi-correlated error structure, we obtain

$$V\left[\hat{\alpha}_{01}\right] = 3\sigma^2/NP;\ 4V\left[\hat{\alpha}_{11}\right] = \left(3\sigma^2/NP\right)[2 - 2\rho];\ 4V\left[\hat{\alpha}_{21}\right] = \left(3\sigma^2/NP\right)[6 - 6\rho].$$

Further,

$$Cov\left[\hat{\alpha}_{01};\ \hat{\alpha}_{11}\right] = 0;\ Cov\left[\hat{\alpha}_{01};\ \hat{\alpha}_{21}\right] = -\left(3\sigma^2/NP\right)[1 - \rho];\ Cov\left[\hat{\alpha}_{11};\ \hat{\alpha}_{21}\right] = 0.$$

Likewise, we have similar results for the other six parameters.

For unbiased prediction, we need an unbiased estimate of $a\alpha_{01} + b\alpha_{02} + c\alpha_{03}$ which is provided by

$$a\bar{y}_{(1,0,0)}(0) + b\bar{y}_{(0,1,0)}(0) + c\,\bar{y}_{(0,0,1)}(0).$$

We may use the above as an unbiased predictor of the yield based on the mixture ($a;\ b;\ c$) at time point 0 and an expression for the variance of the predictor is given by

$$v = 3\sigma^2/N\left[a^2/P + b^2/Q + c^2/R\right].$$

Discussion 5.1 If we confine only to the time point $t=0$ and look at the model expectations of $\bar{y}_{(1,0,0)}(0)$; $\bar{y}_{(0,1,0)}(0)$; $\bar{y}_{(1,0,1)}(0)$, we arrive at the respective means as α_{01}, α_{02}, α_{03}. For the mixture ($a;\ b;\ c$), the model expectation at time point $t = 0$ is $a\alpha_{01} + b\alpha_{02} + c\alpha_{03}$ for which an unbiased estimator is given by $a\bar{y}_{(1,0,0)}(0) + b\bar{y}_{(0,1,0)}(0) + c\,\bar{y}_{(0,0,1)}(0)$. It is easy to ascertain that this estimator has variance given by $3\sigma^2/N\left[a^2/P + b^2/Q + c^2/R\right]$ and this attains its least value of $3\sigma^2/N$ when P

= *a; Q* = *b; R* = *c*. One would wonder if this [trivial] optimal mixture design ξ_0 = [(1; 0; 0) (*P* = *a*); (0; 1; 0) (*Q* = *b*); (0; 0; 1) (*R* = *c*)]—for prediction of the yield based on the mixture (*a; b; c*) at time point 0—is indeed unique of its kind! The claim mentioned below asserts that this is far from being so. There is indeed a large class of such optimal mixture designs. The following result holds true.

Theorem 5.1 *At time t* = *0, optimal prediction of the yield corresponding to the mixture (a; b; c) can be achieved by a class of mixture designs, including in particular the 'trivial' mixture design* ξ_0 *defined above.*

Proof Denote by $\boldsymbol{M}(\xi)$, the moment matrix of the parameters α_{01}, α_{02}, α_{03} based on a general mixture design ξ having three distinct support points. We confine to the subclass of designs of the form ξ: [(*p; q; r*) *P*; (*r; p; q*) *Q*; (*q; r; p*) *R*] subject to $0< p; q; r<1; p + q + r = 1$ and $0<P, Q, R <1; P + Q + R = 1$. It follows that the moment matrix of the model parameters is given by

$$M(\xi) = X\Delta X'.$$

Here, *X* is a matrix of order 3 × 3 with the rows [(*p; r; q*); (*q; p; r*); (*r; q; p*)] and $\boldsymbol{\Delta}$ is the diagonal matrix with elements *P*, *Q*, *R*. It follows that $\det(\boldsymbol{X}) = p^2 + q^2 + r^2 - pq - pr - qr$ and $\det(\boldsymbol{X}\,\boldsymbol{X}^{-1})$ is a matrix having elements in the rows as

$$\left(p^2-qr\right)\left(q^2-pr\right)\left(r^2-pq\right);\ \left(q^2-pr\right)\left(r^2-pq\right)\left(p^2-qr\right);\ \left(r^2-pq\right)\left(p^2-qr\right)\left(q^2-pr\right).$$

It follows that variance of the unbiased predictor for the yield corresponding to the mixture (*a; b; c*) is, apart from a constant, given by $A^2/P + B^2/Q + C^2/R$, where

$$A\ \det(X) = \left(r^2 - pq\right) + a(p - r) + b(q - r);$$
$$B\ \det(X) = \left(q^2 - pr\right) + a(r - q) + b(p - q);$$
$$C\ \det(X) = \left(p^2 - qr\right) + a(q - p) + b(r - p).$$

Note that $A + B + C = 1$. Using Cauchy–Schwartz inequality, we have $A^2/P + B^2/Q + C^2/R \geq 1$. Given (*a; b; c*), if *p*, *q*, *r* are such that *A*, *B*, *C*>0, then we get an optimal design for which the lower bound to variance is attained by ξ_0: [(1; 0; 0) *a*; (0; 1; 0) *b*; (0; 0; 1) *c*]. We now argue that a non-trivial solution exists in the above by taking $q = r$. This means $p + 2q = 1$ and, hence, $\det(\boldsymbol{X}) = (p-q)^2 = (1 - 3q)^2$. We now assume, without any loss of generality, $a \leq b \leq c$ so that $a \leq 1/3$; $c \geq 1/3$. For the choice of the mixing proportions, i.e. (*p; q; q*), etc., we choose $q < a \leq 1/3$ so that $A\,(1 - 3\,q)^2 = (a - q)\,(p - q)$; $B(1 - 3\,q)^2 = (b - q)\,(p - q)$; $C(1 - 3\,q)^2 = (c - q)\,(p - q)$. Hence, for equality in the C–S inequality, our choice of *P*, *Q*, *R* are given by $P = (a-q)/(1 - 3q)$; $Q = (b - q)/(1-3q)$; $R = (c - q)/(1 - 3q)$. One could as well go for a choice of $1/2 > q > c$, provided $c < 1/2$, which would yield $P = (q - a)/(3q - 1)$; $Q = (q - b)/(3q - 1)$; $R = (q - c)/(3q - 1)$.

We illustrate these choices below.

Example 5.1 Consider $a = 1/7$; $b = 2/7$; $c = 4/7$. Start with the mixture $\xi_{00} = [(p;\ q;\ q)\ P^*;\ (q;\ p;\ q)\ Q^*;\ (q;\ q;\ p)\ R^*]$. For any choice of $q < a = 1/7$, we can determine the mixing proportions $P^*; Q^*; R^*$ such that variance of the predicted yield attains the lower bound dictated by an application of the so-called trivial optimal design ξ_0. If $q = 1/8$, such an optimum design is ξ_{00}: [(3/4; 1/8; 1/8) 1/35; (1/8; 3/4; 1/8) 9/35; (1/8; 1/8; 3/4) 25/35].

Example 5.2 Consider $a = 1/5$; $b = 7/20$; $c = 9/20$. Start with the mixture $\xi_{00} = [(p;\ q;\ q)\ P^*;\ (q;\ p;\ q)\ Q^*;\ (q;\ q;\ p)\ R^*]$. For any choice of $q < 1/5$, we can determine the mixing proportions P^*, Q^*, R^* such that the variance of the predicted yield attains the lower bound dictated by an application of ξ_0. If $q = 1/8$, the optimum design is

$$\xi_{00} : [(3/4; 1/8; 1/8)\ 3/25;\ (1/8; 3/4; 1/8)\ 9/25;\ (1/8; 1/8; 3/4)\ 13/25.$$

Also for $q = 12/25 > c = 9/20$, the optimum design is

$$\xi_{00} : [(13/25; 6/25; 6/25)\ 7/11;\ (6/25; 13/25; 6/25)\ 13/44;\ (6/25; 6/25; 13/25)\ 3/44].$$

Example 5.3 Consider $\boldsymbol{x} = (0.4;\ 0.4;\ 0.2)$. Our trial-and-error attempts suggest the existence of non-trivial optimal designs. We cite one such solution:

$$\xi_{00} : [(0.6;\ 0.3;\ 0.1)\ 11/19;\ (0.1;\ 0.6;\ 0.3)\ 7/19;\ (0.3;\ 0.1;\ 0.6)\ 1/19]$$

apart from the trivial optimal design:

$$\xi_0 : [(1, 0, 0)\ 0.4;\ (0, 1, 0)\ 0.4;\ (0, 0, 1)\ 0.2].$$

Example 5.4 Consider $\boldsymbol{x} = (0.5;\ 0.3;\ 0.2)$. Our trial-and-error solution is given by

$$\xi_{00} : [(0.40;\ 0.35;\ 0.25)\ 55/123;\ (0.35;\ 0.25;\ 0.40)\ 63/123;\ (0.25;\ 0.40;\ 0.35)\ 5/123],$$

apart from the trivial optimal prediction design:

$$\xi_0 : [(1;\ 0;\ 0)\ 0.5;\ (0;\ 1;\ 0)\ 0.3;\ (0;\ 0;\ 1)\ 0.2].$$

Discussion 5.2 If we confine only to the time points $t = -1$, 0, 1 and look at the model expectations of $\bar{y}_{(1,0,0)}(0)$; $\bar{y}_{(0,1,0)}(0)$; $\bar{y}_{(1,0,1)}(0)$ at all the three time points, we arrive at the respective means involving the model parameters and their linear functions. From these, we can readily solve for the estimates of all the nine model parameters. We have noted earlier that for the mixture $(a;\ b;\ c)$, the model expectation is $a\alpha_{01} + b\alpha_{02} + c\alpha_{03}$ for time point 0 for which an unbiased estimator is given by $a\,\bar{y}_{(1,0,0)}(0) + b\,\bar{y}_{(0,1,0)}(0) + c\,\bar{y}_{(0,0,1)}(0)$. It was further observed that this estimator has variance given by $3\sigma^2/N\left[a^2/P + b^2/Q + c^2/R\right]$. Likewise, we can work out unbiased predictors of the yields for time points $t = -1; +1$ and also work out their variances.

It turns out that total of variances of all the three predictors is proportional to $A^2/P + B^2/Q + C^2/R$ and this attains its least value when $P = a; Q = b; R = c$. One would wonder if this [trivial] optimal mixture design $\xi_0 = [(1;0;0)\, P = a; (0;1;0)\, Q = b; (0;0;1)\, R = c]$ is indeed unique of its kind when it comes to simultaneous prediction at all the three time points for the given mixture, viz., $(a; b; c)$!

The above study is confined to homogeneous linear mixture model only. We did not venture into homogeneous quadratic mixture model.

In its most general form, a q-component homogeneous p-th degree mixture model could be taken up over the symmetric time interval $[-1, 1]$ and optimality studies could be taken up for both estimation and prediction. It will not be a trivial exercise in case the correlation matrix assumes a very general form.

Acknowledgements The authors thank the anonymous referees for their fruitful comments. Much of the final shape of the manuscript rested on their insightful and helpful suggestions and recommendations towards better presentation of our ideas.

References

Abt, M., Liski, E. P., Mandal, N. K., & Sinha, B. K. (1997). Optimal design in growth curve models: Part I. Correlated model for linear growth: Optimal designs for slope parameter estimation and growth prediction. *Journal of Statistical Planning and Inference, 64,* 141–150.

Abt, M., Gaffke, N., Liski, E. P., & Sinha, B. K. (1998). Optimal designs for growth curve models: Part II. Correlated model for quadratic growth: Optimal designs for slope parameters and growth prediction. *Journal of Statistical Planning and Inference, 67,* 287–296.

Afsarinejad, K. (1990). Repeated measurements designs—a review. *Communication in Statistics—Theory and Methods, 19,* 3985–4028.

Box, G. E. P. (1950). Problem in the analysis of growth and wear curves. *Biometrics, 6,* 362–389.

de la Garza, A. (1954). Spacing of information in polynomial regression. *The Annals of Mathematical Statistics, 25,* 123–130.

Galil, Z., & Kiefer, J. (1977). Comparison of simplex designs for quadratic mixture models. *Technometrics, 19,* 445–453.

Kiefer, J. (1974). General equivalence theory for optimal designs. *The Annals of Statistics, 2,* 849–879.

Potthoff, R. F., & Roy, S. N. (1964). Generalized multivariate analysis of variance model useful especially for growth curve problems. *Biometrika, 51,* 313–326.

Rao, C. R. (1958). Comparison of growth curves. *Biometrics, 14,* 1–16.

Rao, C. R. (1965). Theory of least squares when the parameters are stochastic and its application to the analysis of growth curves. *Biometrika, 52,* 447–458.

Scheffé, H. (1958). Experiments with mixtures. *Journal of Royal Statistical Society, Series B, 20,* 344–360.

Sen, P. K. (2013). Some statistical perspectives of growth curve models in health care plans. In R. Dasgupta (Ed.), *Advances in growth curve models* (Vol. 46, pp. 35–48).New York: Springer Proceedings in Mathematics & Statistics.

Shah, K. R., & Sinha, B. K. (1989) *Theory of optimal designs.* Berlin: Springer.

Sinha, B. K., Mandal, N. K., Pal, M., & Das, P. (2014). *Optimal mixture experiments* (Vol. 1028). Springer-Verlag Lecture Notes in Statistics.

Von Rosen, D. (1991). The growth curve model—a review. *Communication in Statistics—Theory and Methods, 20*(9), 2791–2822.

Tuber Crop Growth Model, Performance Rate, and Some Characterization Theorems

Ratan Dasgupta

Abstract Geometric and exponential distributions may be used for modeling number of tubers and yield of crop. Geometric distribution is discrete version of appropriate exponential distribution and both the distributions have memoryless property. We model a real dataset on number of potato tubers arising from a growth experiment conducted in Giridih farmland and study the properties of these and related distributions in terms of performance rate (Dasgupta 2018) and hazard rate. Some characterization theorems are proved for discrete and continuous random variables.

Keywords Exponential growth model · Tuber crop · Performance rate · Hazard rate

MS Subject Classification: Primary · 62E10 · Secondary · 62P10

1 Introduction

Tuber crop potato (*Solanum tuberosum*) is a staple food in many countries. The number of tubers in plants is related to yield. Number of tubers grown in a plant may be viewed in terms of additional tubers grown from start, sequentially appearing in plants, as this process gradually approaches to the "last tuber" grown at the mature stage of plant. When the probability of an additional tuber grown is independent of the number of tubers already grown, especially when the tuber growth process is fast enough in the period of abundant photosynthesis in plants, then the number of tubers

R. Dasgupta (✉)
Theoretical Statistics and Mathematics Unit, Indian Statistical Institute,
203 B.T. Road, Kolkata 700108, India
e-mail: rdgupta@isical.ac.in; ratandasgupta@gmail.com

R. Dasgupta (ed.), *Advances in Growth Curve and Structural Equation Modeling*,
https://doi.org/10.1007/978-981-13-0980-9_7

in a plant may follow a geometric distribution; this is the only discrete distribution with memoryless property.

If the intensity of tuber growth is plant specific, then a heavy-tailed Pareto-like model is more appropriate for number of tubers aggregated over many plants. As seen in Dasgupta (2011, 2013), mixture of exponential densities with a beta prior induced on exponential of intensity function for variables with memoryless property (viz., exponential random variables) results in a heavy-tailed distribution with polynomial decay, much like a Pareto variable, rather than the original exponential distribution. This mixture distribution is appropriate for modeling variables in situations when memoryless property of a variable may hold only in subgroups, and the task is to model the variable aggregated over many such groups.

However, when the prior is concentrated on a small neighborhood of a parametric point, then the original exponential distribution with memoryless property is not changed much in mixtures, after aggregation of tubers from different plants.

In Sect. 2, we present data of a tuber crop in a growth experiment conducted in Giridih farm of Indian Statistical Institute. The number of tubers in potato plants in the experiment is seen to follow geometric distribution. However, yield of potato may not follow an exponential distribution.

Abundance in crop production depends on the agro-climatic environment. The terminology "performance rate" was introduced in Dasgupta (2018) as an indicator of production system status, especially in the context of industrial productions. The notion is further explained as a substitute of the term "reverse hazard rate."

In Sect. 3, some characterization theorems in discrete and continuous case are proved. Characterization results obtained in terms of constant performance rate for discrete and continuous case lead to distributions with increasing p.m.f./density function till the end of support, and the distributions are of geometric form and exponential form, respectively. These distributions have upper bound on the variate values, unlike geometric and exponential distribution of constant hazard rate. Characterization of distributions, especially in discrete case, with specific form of performance rate and hazard rates including polynomial decay are also investigated.

2 Data from a Growth Experiment and Search for a Model

Memoryless property in discrete random variable leads to geometric distribution. This may be checked in case of production of tuber crops. An observed dataset on number of potato tubers in a growth experiment with 100 potato seeds conducted during 2016–17 is seen to have a satisfactory fit with geometric distribution under χ^2 goodness-of-fit criterion. The same is also corroborated in approximate linear relationship observed in logarithm of sample survival function with number of tubers, a phenomenon expected under the geometric model.

Consider a random variable X with c.d.f. F and density function f. Denote, $\bar{F} = 1 - F$, an almost linear relationship of $\log(n\bar{F}_n(x))$ with x, the number of tubers per plant is seen in the observed data, in a total of $n = 99$ plants, where $F_n(x)$

is the sample c.d.f. See Fig. 1. One seed did not germinate out of 100 seed corms planted.

Geometric distribution is therefore an appropriate model for the data on number of potato tubers; these are obtained from a growth experiment conducted in Indian Statistical Institute Giridih farm during the year 2016–17. One hundred potato seed corm with mean weight of 20 g were planted on December 23, 2016, in 100 pits consisting of four blocks each having 25 pits, of which one seed corm did not germinate, and the crop was harvested on March 25, 2017.

A number of tubers in ascending order from 99 pits are as follows.
1, 1, 1, 1, 1, 1, 2, 2, 2, 2, 2, 2, 2, 2, 2, 2, 2, 2, 2, 3, 3, 3, 3, 3, 3, 3, 3, 3, 3, 3, 4, 4, 4, 4, 4, 4, 4, 5, 5, 5, 5, 5, 5, 5, 5, 5, 5, 5, 5, 5, 6, 6, 6, 6, 6, 6, 6, 6, 6, 6, 7, 7, 7, 7, 7, 7, 7, 8, 8, 8, 8, 9, 9, 10, 10, 10, 10, 11, 11, 11, 11, 11, 11, 12, 14, 16, 17, 17, 18, 18, 18, 20, 22, 24, 24, 26, 26, 34, 40.

Parameter of the geometric distribution

$$P(x) = p(1-p)^{x-1}, 0 < p < 1,\ x = 1, 2, 3, \ldots \tag{1}$$

is estimated as the inverse of the raw mean from the above data. This provides $p = 1/7.757576 = 0.12891$. The value of chi-square of goodness of fit is $\chi^2 = 111.0967 - 99 = 12.0967$ with 7 d.f. See Table 1.

p-value of significance is $p = 0.0974$. The fit seems good, and this is also evident from Fig. 1 of least square fitted line on the data points $(x,\ \log(n\bar{F}_n(x)))$ with intercept $4.547754 \approx \log n = \log 99 = 4.59512$, and with high value of correlation coefficient $r = 0.9922$. Observe that in Fig. 1 we have considered the raw data in full, whereas Table 1 considers grouped data from above raw data to test goodness of fit. The geometric model seems satisfactory in both the occasions.

Yield from potato plants in a pit may be viewed as number of tubers multiplied by weight of a *typical* tuber. In a continuous scale, an appropriate model for yield is then exponential distribution, in view of the proposed geometric distribution for number of tubers. However, in this particular case, exponential fit to yield is not so good.

Potato yield (kg) in ascending order from 99 pits is as follows.

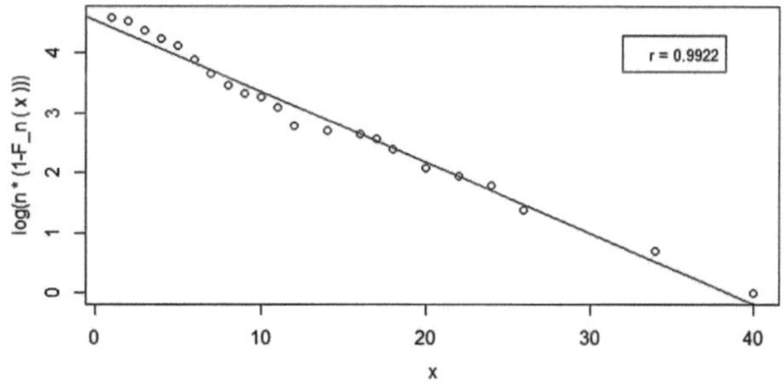

Fig. 1 Model fit for no. of potato (2016–17)

Table 1 Geometric model fit for no. of potato tubers (2016–17)

No. of tuber (i)	Observed frequency (O_i)	c.d.f	e_i	O_i^2/e_i
1	6	0.12890625	12.76171875	2.820936639
2	13	0.241195679	11.11665344	15.20241689
3	11	0.339010298	9.683647335	12.49529189
4	7	0.424216002	8.43536467	5.808877496
5	13	0.498438158	7.347993443	22.99947616
6	10	0.563092614	6.400791164	15.62306869
7	7	0.619412707	5.575689178	8.788151283
8	5	0.668472788	4.856947995	5.147265325
[9-)	27	1	32.82119402	22.21125775
Total	99		99	111.0967421

0.011, 0.014, 0.017, 0.018, 0.021, 0.022, 0.026, 0.026, 0.027, 0.032, 0.041, 0.046, 0.047, 0.049, 0.051, 0.052, 0.055, 0.056, 0.057, 0.059, 0.060, 0.061, 0.061, 0.062, 0.064, 0.065, 0.072, 0.073, 0.079, 0.080, 0.080, 0.080, 0.085, 0.085, 0.085, 0.089, 0.089, 0.093, 0.099, 0.100, 0.101, 0.103, 0.104, 0.105, 0.105, 0.106, 0.107, 0.108, 0.110, 0.115, 0.116, 0.122, 0.122, 0.124, 0.125, 0.130, 0.132, 0.133, 0.139, 0.144, 0.144, 0.146, 0.147, 0.149, 0.150, 0.151, 0.153, 0.155, 0.155, 0.155, 0.157, 0.160, 0.170, 0.172, 0.174, 0.174, 0.174, 0.180, 0.180, 0.183, 0.184, 0.185, 0.185, 0.196, 0.205, 0.209, 0.213, 0.217, 0.220, 0.222, 0.244, 0.256, 0.282, 0.286, 0.300, 0.321, 0.360, 0.365, 0.410.

Exponential model fit on tuber yield x based on $(x,\ \log(n\bar{F}_n(x)))$ for the above 99 raw data is shown in Fig. 2. Estimate of intensity from raw data is $\lambda = 1/\bar{x} = 1/0.12869 = 7.7706$.

For the grouped data with 10 class intervals shown in Table 2, the value of intensity is a little bit low; $\lambda = 1/\bar{x} = 1/0.129869 = 7.7001$.

Chi-square of goodness of fit in grouped data is $\chi^2 = 125.7817216 - 99 = 26.7817$, with $d.f = 8$. With intensity estimated from raw data and when same class intervals are considered for goodness of fit, chi-square is higher than before, $\chi^2 = 42.6459$, with $d.f = 8$. The values are high possibly due to the fact that the

Fig. 2 Model fit for potato yield (2016–17)

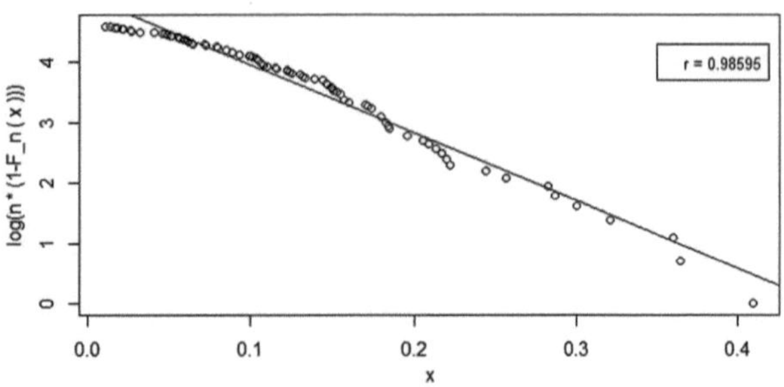

underlying assumption made that yield equals number of tubers multiplied by weight of a typical tuber, may not be adequate for the entire range of tuber varieties, from large to small in sizes, but may be an appropriate assumption only for tubers of homogenous size. For a similar situation with data on bulb crop garlic, see Dasgupta (2017).

3 Performance Rate and Some Characterization Theorems

Memoryless distributions have application in lifetime analysis, e.g., in quality assurance of industrial products. Performance rate is a newly introduced terminology for the quantity $\frac{f(t)}{F(t)}$ (Dasgupta 2018), in place of earlier term "reverse hazard rate" adopted in relation to time reversal or change in sign of the variable. The quantity $\epsilon \frac{f(t)}{F(t)}$ represents the probability that the system fails in a ϵ small neighborhood $(t-\epsilon, t]$, $\epsilon > 0$ given that the lifetime $T \leq t$. High values of $\frac{f(t)}{F(t)}$ indicate satisfactory performance of the system, as the system worked till the time near about t, while lifetime is restricted up to t; thus the measure performance rate provides a meaningful glance into the system performance, with the interpretation: the higher the rate, the better the performance.

The terminology reverse hazard rate on the contrary does not convey any meaning of the system status in comparison with the present terminology *performance rate.*

Exponential distribution is the continuous counterpart of geometric distribution with constant hazard rate.

Characterizations of geometric and exponential distribution in terms of independence of minimum and difference of two variables are investigated in Crawford

Table 2 Exponential model fit for potato yield (kg) (2016–17)

Class interval	Observed frequency (O_i)	Theo_cum_freq	e_i	O_i^2/e_i
(0.011–0.027]	9	13.5800572	13.5800572	5.964628779
(0.027–0.059]	10	28.10407484	14.52401765	6.885147239
(0.059–0.08]	10	41.2898464	13.18577155	7.583932393
(0.08–0.1]	10	49.78271115	8.49286475	11.77458996
(0.1–0.115]	10	56.0364268	6.253715656	15.99049357
(0.115–0.144]	10	62.78343765	6.747010844	14.82137828
(0.144–0.157]	11	68.23290636	5.449468719	22.20399937
(0.157–0.183]	9	72.55617061	4.323264243	18.73584298
(0.183–0.222]	10	78.45430664	5.898136032	16.95450893
(0.222-]	10	90.48957006	20.54569336	4.867200062
Total	99		99	125.7817216

(1966). Below we characterize distributions with constant and positive performance rate in limit.

Note that $f(t) = kF(t) = k\int_{-\infty}^{t} f(x)dx$ implies $f'(t) = kf(t)$, when f' exists. The later is equivalent to $f(t) = e^{kt+c}$ on the support of f. The c.d.f. of the distribution is therefore of the form $F(t) = e^{k(t-m)}$, with constant performance rate $k > 0$ and the range of the variable is $t \in (-\infty, m]$.

The result can be extended if the performance rate $k(t) = k + o(1)$ is a constant to a first degree of approximation, $o(1)$ term is negligible for t large. $f(t) = k(t)F(t)$ may be written as $\frac{d}{dt}\log F(t) = k(t)$, and this leads the functional form of the distribution function when $k(t)$ is constant with small oscillations. The form of the c.d.f. is then $\log F(t) = k(t-m)(1+o(1))$, contribution from the $o(1)$ term to the integration over the range $(-\infty, t]$ is $o(t)$, which is negligible compared to the leading term for t large.

Estimated performance rate may have small oscillation around the true value due to sampling error. When allowance is made for that in computing F, the above provides an estimate of error, viz., $o(t)$. This is slightly extended version of Proposition 2.1 of Block et al. (1998).

The distribution is of exponential form with positive coefficient in the exponent, and therefore has a finite upper end point, unlike exponential distribution of decreasing density and constant hazard rate on the range $(0, \infty)$.

Performance rate and hazard rate are equal at the median of a distribution. To obtain equality of these two rates in general case, consider a sequence of normal distributions $N(\mu_m, 1/n)$ with decreasing variance, where $\mu_m \in Q$, the set of rational numbers. Linear combination of μ_m spans R. The sequence of normal distributions and linear combinations of these distributions weakly converge to the Lebesgue measure in R, as $n \to \infty$. The performance rate and hazard rate are equal at μ_m and linear combinations of it, spanning R.

Improper priors like Lebesgue measure are relevant in Bayesian analysis.

For discrete distributions on $\mathbf{N} = \{0, 1, 2, 3, \ldots\}$ with associated probabilities $p_0, p_1, p_2, p_3, \ldots$, performance rate is constant $k \in (0, 1)$ which implies $p_n / \sum_{i=0}^{n} p_i = k$. This in turn provides the relation $\frac{p_1}{p_0+p_1} = k$, thus $p_1 = \frac{kp_0}{1-k}$; $\frac{p_2}{p_0+p_1+p_2} = k$, thus $p_2 = p_1 + kp_2 = \frac{p_1}{1-k} = \frac{kp_0}{(1-k)^2}$, etc. Thus,

$$p_i = \frac{p_{i-1}}{1-k} = \frac{p_{i-2}}{(1-k)^2} = \cdots = \frac{kp_0}{(1-k)^i} \tag{2}$$

which is increasing in i. This indicates that there is an upper bound on the support of the distribution when performance rate is constant in discrete case. Let the upper bound be n. The total probability being 1, this specifies the initial value p_0 in terms of the performance rate k as $p_0 = \frac{(1-k)}{(1-k)+\{1-(1-k)^n\}}$.

Note that $\frac{p_i}{p_{i+1}} = (1-k) \in (0, 1)$.

Over the state space $\mathbf{N} = \{0, 1, 2, 3, \ldots\}$, consider the distribution $p_k = \frac{1}{(k+1)(k+2)}$, $k \in \mathbf{N}$. This distribution culminates from the uniform prior on the parametric space $\theta \in (0, 1)$ of geometric distributions, see Dasgupta (2011) pp. 506–507, in connection with applications to lifestyle data. We briefly recapitulate the background in Bayesian setup.

The number of successes preceding mth failure with constant probability θ of success at each trial in a sequence of independent trials leads to the negative binomial distribution.

$$P(X = k) = \binom{m+k-1}{m-1}\theta^k(1-\theta)^m \quad k = 0, 1, 2, 3, \ldots \tag{3}$$

In the special case $m = 1$, one gets the geometric distribution for the number of k successes preceding first failure.

$$P(X = k) = \theta^k(1-\theta) \quad k = 0, 1, 2, 3, \ldots \tag{4}$$

In our earlier notation, $\theta = 1 - p$. Here, the probabilities decrease exponentially fast as k increases. However, if the probability of occurrence in a single trial is not constant but follows a prior distribution

$$f_{\alpha,\beta}(\theta) = \frac{\Gamma(\alpha+\beta)}{\Gamma(\alpha)\Gamma(\beta)}\theta^{\alpha-1}(1-\theta)^{\beta-1}, \ \alpha > 0, \ \beta > 0; \ \theta \in (0, 1) \tag{5}$$

then the resultant probabilities decrease polynomially fast as k increases, resulting in a distribution with thick tail. This may happen when several groups of geometric variables with different intensities are aggregated. The marginal distribution of X is then

$$P(X = k) = \frac{\Gamma(\alpha+\beta)}{\Gamma(\alpha)\Gamma(\beta)} \int_0^1 \theta^{\alpha+k-1}(1-\theta)^{\beta} d\theta$$

$$= \frac{\Gamma(\alpha+\beta)}{\Gamma(\alpha)\Gamma(\beta)} \frac{\Gamma(\alpha+k)\Gamma(\beta+1)}{\Gamma(\alpha+k+\beta+1)} = O_e((k+1)^{-(\beta+1)}) \quad k = 0, 1, 2, 3, \ldots \tag{6}$$

where O_e represents the exact order. The result is obtained from Stirling's approximation. The ratio of successive probabilities is monotonically decreasing,

$$P(X = k+1)/P(X = k) = (1 + \frac{\beta+1}{k+\alpha})^{-1} \approx (1 - \frac{\beta+1}{k+\alpha}), \ \beta > 0 \tag{7}$$

as $k \to \infty$.

In the special case of noninformative uniform prior over (0, 1), one has $\alpha = \beta = 1$ and the probability in (6) reduces to

$$p_k = P(X = k) = \frac{1}{(k+1)(k+2)} = \frac{1}{(k+1)} - \frac{1}{(k+2)}, \ k = 0, 1, 2, 3, \ldots \tag{8}$$

The probabilities can be expressed in terms a general harmonic sequence of monotonic constants, thus enabling computation of survival function easy. Such sequence of constants is also relevant in derivation of Gomphertz model, see pp. 192–193 of Dasgupta (2015). Hazard rate of the discrete distribution (8) is $h_k = \frac{p_k}{\sum_{i=k}^{\infty} p_i} = \frac{1}{k+2}$, which is inversely decreasing as k increases. This form of h characterizes the distribution.

To see this note that $h_0 = \frac{p_0}{\sum_{i=0}^{\infty} p_i} = \frac{1}{2}$ states $p_0 = \frac{1}{2}$, $h_1 = \frac{p_1}{\sum_{i=1}^{\infty} p_i} = \frac{1}{3}$ implies $\frac{p_1}{1-p_0} = \frac{1}{3}$, i.e., $p_1 = \frac{1}{6}, \ldots$ Assume that the result holds up to h_{k-1}, and we prove the result by induction. Then $h_k = \frac{p_k}{1-\sum_{i=0}^{k-1} p_i} = p_k/(k+1)^{-1}$, proving the result.

Thus we have the following three characterization theorems.

Theorem 1 *Let the continuous random variable X have performance rate* $k(t) = f(t)/F(t)$. *The rate k is constant and positive iff the distribution function is of the form* $F(t) = e^{k(t-m)}$, *for* $t \in (-\infty, m]$.

The performance rate $k(t) = k + o(1)$ *is a constant to a first degree of approximation iff the density function is of the form* $f(t) = ke^{k(t-m)(1+o(1))}$, *for* $t \in (-\infty, m]$.

For the improper prior Lebesgue measure on the real line R, performance rate and hazard rate may be interpreted as equal.

Theorem 2 *For discrete distribution on set of integers* $\mathbf{N} = \{0, 1, 2, 3, \ldots\}$ *with associated probabilities* $p_0, p_1, p_2, p_3, \ldots$, *performance rate* $k(i) = p_i / \sum_{j=0}^{i} p_j$ *is constant* $k \in (0, 1)$ *iff the p.m.f.* p_i *is of the form* $p_i = \frac{kp_0}{(1-k)^i}$, *for* $i = 1, 2, 3, \ldots, n$ *where* $p_0 = \frac{(1-k)}{(1-k)+\{1-(1-k)^n\}}$.

The above distribution is of geometric type and has an upper bound. The ratio $(1-k)^{-1}$ *of the consecutive probabilities is greater than 1. Mode of the distribution is at the largest end point. In the present setup, the variate values are equispaced. Characterization theorems where the values are not necessarily equispaced are considered in Dasgupta (1993) in an industrial context for the characteristic eccentricity.*

Theorem 3 *Consider discrete distributions on set of integers* $\mathbf{N} = \{0, 1, 2, 3, \ldots\}$. *The hazard rate is inversely decreasing and of the form* $h_k = \frac{p_k}{\sum_{i=k}^{\infty} p_i} = \frac{1}{k+2}$ *iff the distribution is* $p_k = \frac{1}{(k+1)(k+2)}$, $k \in \mathbf{N}$.

The next theorem states the form of the discrete distribution when the form of the performance rate is of the type of hazard rate stated in Theorem 3. In Dasgupta (2018), it is shown that extended uniform distribution, and thus uniform distribution, has decreasing performance rate $O(1/x)$. *When the performance rate is of same form in discrete case, then it leads to discrete uniform distribution, as the following result states.*

Theorem 4 *Consider discrete distribution on a finite set of integers* $A = \{0, 1, 2, 3, \ldots, n\}$. *Then, the performance rate* $\frac{p_i}{p_0+p_1+p_2\cdots+p_i} = k(i) = \frac{1}{i+1}$, $\forall i \in A$ *iff the distribution is uniform on A.*

Proof One way implication is obvious. We show the "only if" part. $\frac{p_1}{p_0+p_1} = k(1) = 1/2$, thus $p_1 = p_0$; $\frac{p_2}{p_0+p_1+p_2} = k(2) = 1/3$, thus $2p_2 = p_0 + p_1$, i.e., $p_2 = p_0$; etc. Now assume $p_0 = p_1 = p_2 = \cdots = p_i$. Then, $\frac{p_{i+1}}{p_0+p_1+p_2+\cdots+p_i+p_{i+1}} = k(i+1) = \frac{1}{i+2}$ implies $\frac{p_{i+1}}{(i+1)p_0+p_{i+1}} = \frac{1}{i+2}$, i.e., $p_{i+1} = p_0$, for $i + 1 \in A$.

References

Block, H., Savits, T., & Singh, H. (1998). The reversed hazard rate function. *Probability in the Engineering and Informational Sciences*, *12*, 69–90.

Crawford, G. B. (1966). Characterization of geometric and exponential distributions. *The Annals of Mathematical Statistics*, *37*, 1790–1795.

Dasgupta, R. (1993). Cauchy equation on discrete domain and some characterization theorems. *Theory of Probability and Its Applications*, *38*(3), 520–524.

Dasgupta, R. (2011). Discrete distributions with application to lifestyle data. In: *International conference on productivity, quality, reliability, optimization and modeling proceedings* (Vol. 1, pp. 502–520). New Delhi: Allied Publishers.

Dasgupta, R. (2013). Tuber crop growth and pareto model. In *advances in growth curve models: topics from the Indian statistical institute. Springer proceedings in mathematics & statistics* (Vol. 46, pp. 185–198) Chapter 10.

Dasgupta R. (2015). Growth model of some vernacular word usage during political transition. In *Growth curve and structural equation modeling: Topics from the Indian Statistical Institute* (pp. 171–193) Chapter 10. Springer.

Dasgupta, R. (2017). Model selection and validation in agricultural context: Extended uniform distribution and some characterization theorems. In *Growth curve models and applications* (pp. 183–198) Chapter 9. Springer.

Dasgupta, R. (2018). *Characterization of extended uniform distribution and its applications, advances in growth curve and structural equation modeling: Proceedings 2017 Springer (USA)*, Appearing in this volume as chapter 3.

An Overview of Growth and Malnutrition Among 5–18-Year Children in India

Susmita Bharati, Manoranjan Pal and Premananda Bharati

Abstract This paper comprises of assessment of growth and nutritional status of 5–18-year children in India along with finding the key determinants of undernutrition and overnutrition. For this study, the data were collected from the second round of Indian Human Development Survey (IHDS-II) of 2011–12. Growth study has been done through height and weight and nutritional assessment through age-and sex-specific BMI. Data on socio-economic background are gender, religion and per capita expenditure, which were taken to see the effect of these variables on growth and nutritional status.The results show that growth of children is continuous and ascending in nature. However, a sudden influx may occur at some point during the ages of 10–14 years among children except for height in girls. This influx may occur due to the onset of adolescent growth spurt. Occurrences of undernutrition are directly related with age but maximum undernutrition occurs between 10 and 15 for boys and 10 and 13 for girls. Analysis of possible regional and socio-economic factors with child malnutrition outcomes does not reveal any substantive causal relations except for per capita expenditure, which has a negative impact on undernutrition and positive impact on obesity.

Keywords Growth status · Malnutrition · Underweight · Overweight · Obese · Socio-economic condition

S. Bharati
Sociological Research Unit, Indian Statistical Institute, 203 B.T. Road, Kolkata 700108, India

M. Pal
Economic Research Unit, Indian Statistical Institute, 203 B.T. Road, Kolkata 700108, India

P. Bharati (✉)
Biological Anthropology Unit, Indian Statistical Institute, 203 B.T. Road, Kolkata 700108, India
e-mail: pbharati@gmail.com

R. Dasgupta (ed.), *Advances in Growth Curve and Structural Equation Modeling*,
https://doi.org/10.1007/978-981-13-0980-9_8

1 Introduction

Malnutrition may cause retardation of growth of a child. Thus, a proper assessment of growth and development of child is necessary to find the state of health of a child. Malnutrition not only means less of nutrition from normal level, but also means excess nutrition as reflected by overweight or obesity. Anthropometric measures are recognized as the best useful techniques to assess the growth and nutritional status of individual or population (Gorstein et al. 1994; Jelliffe 1966; Rao et al. 1986; Borooah 2005), because anthropometric measures are highly sensitive to undernutrition (Martorell and Ho 1984).

The coexistence of undernutrition and 'overweight or obesity' are nothing but forms of malnutrition denoting either lack or excess in energy or nutrients like minerals and vitamins, which are necessary to maintain good health (FAO 2006). And very often, the existence of underweight and overweight people within the same population or same household is very often termed as dual burden in the population or household. Dual burden of malnutrition poses a significant challenge for nutrition interventions for improving health outcomes at both ends of the nutrition spectrum (Doak et al. 2000). The dual burden among households is a recent phenomenon and is most prevalent in middle-income countries such as Kyrgyzstan, Indonesia, Russia, where 9–13% households contain both underweight and overweight members (Doak et al. 2005). Dual burden is less common in high-income countries because in the high-income countries, occurrences of obesity are very likely and occurrence of undernutrition is rare. The emergence of the dual burden household is a result of the epidemiological and nutrition transition—decrease in physical activity levels and a shift in diets to more fats and sugars (Barry and Penny 2004). With the globalization of food markets, developing countries have larger quantities of low cost, high calories foods influencing poorer household to purchase more of these cheap foods (Caballero 2005). The causes of dual burden can be explained by the theory of nutrition transition (Popkin 1993). According to him, nutrition transition refers to a change in the composition of diet, accompanied by changes in the physical activity levels and classified into three stages: receding famine, degenerative disease, and behavioural changes. During transition, the food habit of people is shifted from homemade plant-based food to animal-based food, fat, sugar and processed food. In the third stage, people restart to abstaining from saturated-fat-based and processed foods. And in course of this process, automatically physical labour for normal livelihood decreased with the mechanization of lifestyle and finally non-obligatory physical activity becomes more common as behavioural changes introduces non-communicable disease. As a result, these changes are reflected into body weight, body composition and morbidity. Analysis of the dietary pattern suggests that developing countries are mostly in the first or in the second stage of nutrition transition and are facing dual burden of malnutrition (Kennedy et al. 2006).

Like other developing countries, India is passing on nutrition transition as a beneficiary of 20-year economic reforms, and as a result, huge socio-economic and demographic heterogeneity occurs in different states because transition is not occurring homogenously. Due to differentiation of socio-economy in different states and due to differentiation of Government policy, public distribution system and ICDS in different states, the levels of malnutrition differ from state to state and region to region.

In India, half of the infant deaths are still related to undernutrition. Infant mortality is still very high. Undernutrition and disease are the principal culprits. In Global Hunger Index, India ranks 67 out of 84 countries surveyed. On the other hand, globally childhood obesity has been increasing for the past 3 or 4 decades and India is no exception (Raj et al. 2007; Laxmaiah et al. 2007; Subramanyam et al. 2003; Chhatwal et al. 2004; Khadilkar and Khadilkar 2004; Panjikkaran and Kumari 2009). This coexistence of malnutrition and morbid obesity exemplifies the extreme inequalities of wealth and income in India.

After childhood, adolescence is a time of rapid developmental changes. Adequate nutrition during this period is a fundamental cornerstone of an individual's health. Undernourished adolescents are more likely to face short stature and lower resistance to infections that, in turn, leads to higher risk of morbidity and mortality throughout their lives (Mak and Tan 2012; Coly et al. 2006). Overweight and obese adolescents are at risk factors of developing non-communicable diseases and also suffering from psychological problems such as stigmatization and poor self-esteem (Dietz 2001; Reilly et al. 2003).

The coexistence of under- and overnutrition is not very uncommon in recent times. It may exist within a section of people, e.g., within a religion, region, caste, income group, etc. Since it is not desirable to have either undernourished or over-nourished people, the coexistence of both calls for much concern. The causes, thought to affect the status of health leading to undernutrition or overnutrition and common to all people in the section, should equally affect each people and thus may be eliminated at first as causes of dual burden. The dual burden of malnutrition within the household, i.e. both parents are overweight but children are underweight, or brother underweight but sister overweight and vice versa, calls for greater concern for the same reason.

It is necessary to understand the problems and magnitude of the status of underweight and overweight among schoolgoing 5–18-year children in India. The paper, thus, explores nature and dual burden of malnutrition by states and by different socio-economic categories in India.

2 Data Source

Data have been obtained from Indian Human Development Survey–II (IHDS–II) conducted in 2011–12. The survey was coordinated by the National Council of Applied Economics Research (NCAER), New Delhi and by the University of Maryland. The Indian Human Development Survey-II is a multi-topic survey of 42,152 households in 1503 villages and 971 urban neighbourhoods across India. In this study, data were collected on health, education, employment, economic status, marital status, fertility, gender relations, and social capital on three datasets' basis like individual, household and eligible women. In our study, the sample size is 43,065 of 5–18-year children of both sexes in all the states and union territories of India. Anthropometric variables, like weight and height of the children, have been considered for growth study. The nutritional statuses of children are assessed through BMI for age. BMI for age is used to classify each child into different nutritional statuses like underweight, normal, overweight and obese for each age and gender. Age- and gender-specific cut-points as per CDC (Center for Disease Control) are 85th percentile for the age and sex as a reference point for overweight and the 95th percentile for obesity, on the other side, below 15th percentile is designated as undernourished (WHO 2006). For nutritional assessment, age group of the children have been grouped into three groups such as 5–9, 10–15 and 16–18 years due to obvious body requirements of children in different phases of growth such as juvenile, adolescent and post-adolescent phase. The independent variables are zones, sex, religion and per capita expenditure. India is divided into six zones followed by NFHS classification. Religion is grouped into four categories like Hindu, Muslim, Christian and others, which include the rest of the groups. Per capita expenditure has been grouped into six categories with almost equal frequency. Means and standard deviations (sd) of height and weight of children of age 5–18 years have been calculated by age group and sex. The increment of height and weight between the two successive age groups and by sex have been done using 't' test. This test is usually performed to see whether the means of a variable are equal between two groups. The test performs best when the variable follows normal distribution for both the groups. However, because of the central limit theorem, it can be applied in a reasonably large sample of size greater than 30, say, even if it does not follow normal distribution. The sample size for us is much more than 30 whenever this test is applied in this paper. To find the relation between socio-economic variables and malnutrition of children we have used cross-frequency table to carry out chi-square test. The expected numbers of cases should be greater than or equal to 5 for this test to be applied successfully. This criterion is fulfilled in our cases. Relationship between socio-economic variables with malnutrition was also assessed through categorical logistic regressions. The nutritional status of children has been considered as dependent variable. Here dependent variables are binary. Regression analysis was done by two ways—(i) for the assessment of undernutrition, undernutrition is coded as '1' and the rest are coded as '0'. In the same way, for assessment of overnutrition, overnutrition is coded as '1' and the rest are coded as '0'. Thus, the estimated odd ratio greater than 1 means that there is a positive relation between the variable with the dependent variable. In other words, the increase in the value of the variable makes the chance of becoming overweight more. The opposite is the case when the odd ratio is less than 1. We have used the Statistical Package for Social Sciences version 18 (SPSS-18) for most of the calculations. Significance of level of $p = 0.05$ is considered for all statistical tests.

3 Limitations of the Study

It is necessary to mention that the sample sizes in our study are not same as in the published data because of the following reasons:

The height and weight data are taken only for the cases, where other relevant data are available. We have also excluded some outlying observations to maintain the parity between height and weight.

4 Results

The mean and sd. of height and weight of 5–18-year children in India by age and sex are shown in Tables 1 and 2, Figs. 1 and 2. There is a steady increase in height and weight and no marked gender differences are observed up to 13 years of age, but after 13 years, the magnitude of increment is much greater among boys than girls. At 18 years, the difference between boys and girls is more than 10 cm. in height and 6 kg. in weight. The increment of height at the age of 5 years is the highest and then the increment gradually decreases up to the age of 9 for both boys and girls. After this, it increases (Table 1). This rise of growth during this period may be due to juvenile growth spurt among both boys and girls. After that, among boys, the maximum increment is observed from 12 to 13 and 13 to 14, and this may be due to the onset of adolescent growth spurt among boys. Among girls, the highest rate of increment is observed from 10 to 11 years and 11 to 12 years (Table 1), though not prominent enough. So, among girls, the onset of adolescent growth spurt in height is happening at the age of 10–12 years.

In case of weight, among boys, the highest increment rate is observed from 12 to 13, 13 to 14 and 14 to 15 years. So it may happen due to the onset of adolescent growth spurt during these age groups. In case of girls, high rates of increment are observed in the age groups of 10–14 years (Table 2). So it can be said that these happen due to the onset of adolescent growth spurt.

Table 3 and Fig. 3 describe the nutritional status of 5–18-year children in India by age and sex. It shows the picture of different nutritional grades such as underweight, normal, overweight, obese and overweight or obese of 5–18-year Indian children. It is noticeable that undernutrition is increasing with the age up to age 11 or 12 years. A sudden increase of undernutrition occurs at the age of 7 for both boys and girls and then it keeps rising for few more years. It is also seen that among boys, highest undernutrition is observed from 12 to 15 years, magnitude remains very high. So it may be due to the supply of energy for adolescent growth spurt. Among girls, maximum percentages of undernutrition are noticed from age groups 10–12. So, this may probably happen due to supplementation of energy for the onset of adolescent spurt in these years. Percentage of overweight or obesity has more or less a decreasing pattern with age for all children. In the age group of 5 of both boys and girls, percentages of overweight or obesity are 28.7% and 21.6%, respectively, but at the of age of 18, the percentages for the same come down to only 4.5 and 5.2, respectively.

Table 1 Mean height of 5–18-year children by age and gender differences in India

Age groups (yrs.)	Boys					Girls				
	N	Mean	sd	Increment	t-test	N	Mean	sd	Increment	t-test
5	1540	102.37	9.97	–	–	1326	101.18	9.97	–	–
6	1494	108.98	10.24	6.61	18.03	1335	108.29	9.64	7.11	18.69
7	1723	115.04	10.32	6.06	16.64	1604	114.09	10.09	5.80	15.81
8	1510	121.09	9.97	6.05	16.88	1399	119.43	10.88	5.34	13.95
9	1423	125.22	10.54	4.13	10.91	1301	124.21	10.70	4.78	11.49
10	1836	129.70	10.51	4.48	12.03	1722	129.21	10.69	5.00	12.72
11	1451	135.18	10.97	5.48	14.56	1277	134.68	11.15	5.47	13.61
12	2007	139.47	11.05	4.29	11.29	1857	140.11	10.61	5.43	13.77
13	1531	145.88	11.14	6.41	17.05	1580	144.80	9.66	4.69	13.47
14	1715	152.34	11.31	6.46	16.34	1688	148.36	8.64	3.56	11.13
15	1424	157.49	9.80	5.15	13.49	1549	150.44	7.82	2.08	7.14
16	1495	160.56	10.10	3.07	8.33	1599	151.16	7.14	1.16	2.70
17	1432	162.69	9.00	2.13	5.99	1535	152.22	7.20	1.06	4.16
18	1212	163.43	8.82	0.74	2.15	1501	**152.06**	**6.96**	**−0.16**	**0.63**

All the tests, except for 18-year-old girls, are significant at 5% level

Table 2 Mean weight of 5–18-year total children and by gender differences in India

Age groups (yrs.)	N	Mean	sd	Increment	't' test	N	Mean	sd	Increment	't' test
5	1540	15.66	3.12	–	–	1326	15.17	3.48	–	–
6	1494	17.65	3.68	1.99	16.09**	1335	17.18	3.57	2.01	14.67**
7	1723	19.79	4.34	2.14	14.98**	1604	19.13	4.21	1.95	13.39**
8	1510	22.03	4.87	2.24	13.80**	1399	21.21	4.72	2.08	12.77**
9	1423	24.03	5.46	2.00	10.48**	1301	23.38	5.11	2.17	11.49**
10	1836	26.00	5.84	1.97	9.82**	1722	25.97	6.21	2.59	12.21**
11	1451	29.02	6.84	3.02	13.67**	1277	29.05	7.25	3.08	12.51**
12	2007	31.50	7.19	2.48	10.19**	1857	32.61	7.37	3.56	13.34**
13	1531	35.83	8.40	4.33	16.48**	1580	36.42	7.59	3.81	14.91**
14	1715	40.35	8.95	4.52	14.77**	1688	40.19	7.63	3.77	14.16**
15	1424	44.74	8.82	4.39	13.75**	1549	41.94	7.06	1.75	6.74**
16	1495	47.96	8.97	3.22	9.77**	1599	43.64	6.89	1.70	6.85**
17	1432	50.54	8.68	2.58	7.92**	1535	44.66	7.20	1.02	4.04**
18	**1212**	**51.82**	**8.67**	**1.28**	**3.77****	**1501**	**45.07**	**7.04**	**0.41**	**1.56**

All the tests, except for 18-year-old girls, are significant at 5% level

Table 3 Percentage of nutritional status among 5–18-year children in India by age and sex

Age (yrs.)	Boys						Girls					
	N	Undernutrition	Normal	Overweight	Obese	Overweight or obese	N	Undernutrition	Normal	Overweight	Obese	Overweight or obese
5	1540	25.6	57.8	4.5	12.1	16.6	1326	21.0	65.8	4.6	8.5	13.1
6	1494	25.2	61.2	4.0	9.6	13.7	1335	24.9	63.2	4.4	7.5	11.9
7	1723	31.4	55.7	3.5	9.5	12.9	1604	27.1	62.7	3.2	7.1	10.3
8	1510	32.6	55.9	3.4	8.1	11.5	1399	27.7	62.8	4.3	5.2	9.5
9	1423	35.1	53.1	3.6	8.0	11.8	1301	30.9	59.3	3.8	6.0	9.8
10	1835	38.7	51.3	3.9	6.0	10.0	1722	34.7	56.4	3.9	4.0	8.9
11	1451	36.0	54.7	4.3	5.0	9.3	1277	36.7	55.4	3.8	4.2	7.9
12	2007	42.0	50.0	4.3	3.7	8.1	1857	35.4	57.5	4.6	2.5	7.1
13	1531	39.1	53.3	5.3	2.4	7.6	1580	31.1	63.7	3.2	2.0	5.2
14	1715	39.4	54.9	3.9	1.8	5.7	1688	25.8	68.3	3.8	2.1	5.9
15	1424	38.9	55.8	3.9	1.4	5.3	1549	24.3	71.7	3.0	1.0	4.1
16	1495	36.2	58.5	4.0	1.3	5.4	1599	20.2	75.9	2.9	1.1	3.9
17	1432	33.9	61.4	3.1	1.5	4.7	1535	21.8	74.7	1.8	1.7	3.5
18	1212	35.6	60.9	2.5	1.0	3.5	1501	17.7	77.8	2.7	1.7	4.5

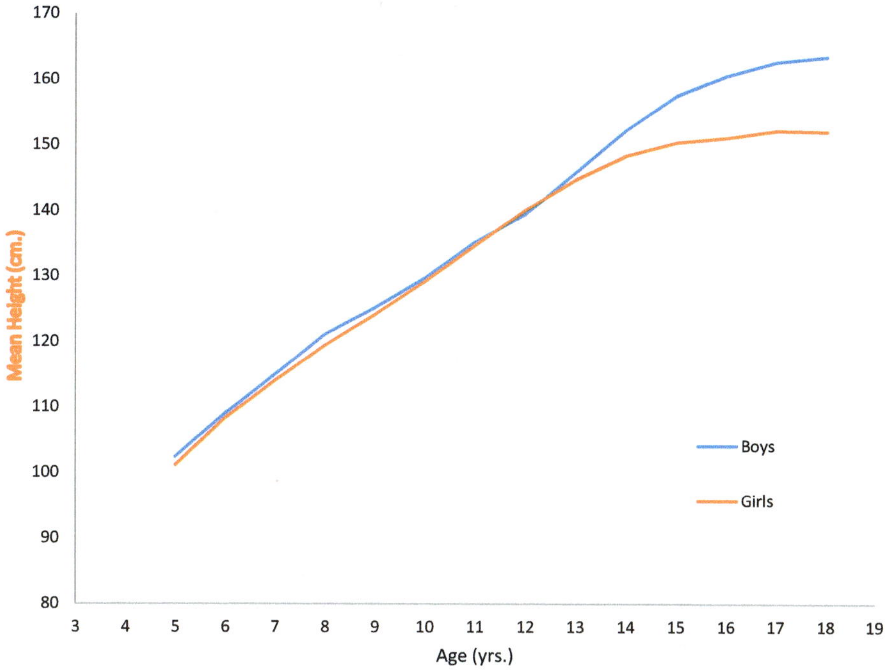

Fig. 1 Distance curve on mean growth of height of 5–18 years children in India

Table 4 represents the percentage of undernutrition and overnutrition in six zones of India of 5–18 years among boys. Some peculiarities are observed regarding the percentages of undernutrition and overnutrition in Northeast zone. The rate of undernutrition among boys is markedly less than other zones. Observe that the sample sizes are very less for Northeast India. Except for South zone, the rate of undernutrition has dramatically increased from 7 years in all zones, except for Northeast zone where from 13 years onwards, the percentage of undernutrition has decreased markedly up to 18 years. Regarding overweight or obesity, in Northeast zone, in the age of 5 and 6 years, very high percentage (62–66%) of overweight or obesity are observed. The percentages of overweight or obese are high up to the age 13 and these percentages fall after the age of 13 till 18 years for all zones. It is also noticeable that in Central zone, percentages of overweight or obese boys are very less.

From Table 5, it is seen that among girls, the percentage of underweight is less in Northeast zone than other zones in all the age groups. And percentages of overweight or obese children are too high to be compared with other zones of India. Though South zone ranks second highest zone but by magnitude, it is far below compared to Northeast zone. In Central and North zones, overweight or obese children are very low than other zones of India. Besides these trends, Tables 4 and 5 represent indirectly the dual burden of malnutrition among 5–18-year children. The peculiarity of Northeast

Table 4 Percentage of under- and overnutrition among 5–18-year boys in different zones of India

Age (yrs.)	Northeast			East			Central			West			North			South		
	N	UW	OW/Obese	N	UW	OW/Obese	N	UW	OW/Obese	N	UW	OW/Obese	N	UW	OW/Obese	N	UW	OW/Obese
5	37	**13.5**	62.2	278	25.2	21.2	439	27.8	9.6	175	29.7	19.4	427	24.6	11.9	184	21.7	25.5
6	30	13.3	66.7	278	23.7	16.9	442	27.4	5.2	166	32.5	20.5	405	20.5	11.6	173	27.7	19.1
7	34	20.6	35.3	323	31.9	13.6	499	31.1	6.4	173	38.2	18.5	492	30.7	11.2	202	29.2	23.8
8	37	21.6	45.9	247	25.9	13.8	458	33.2	2.8	183	44.8	18.6	376	32.4	9.3	209	30.6	19.6
9	33	12.6	36.4	260	32.7	13.5	373	42.9	3.2	164	36.6	22.0	396	32.8	7.8	197	31.0	21.3
10	55	16.4	45.5	345	35.7	11.3	574	42.5	2.3	189	40.7	20.1	456	40.6	7.0	216	33.8	16.7
11	43	27.9	25.6	229	28.8	7.9	359	45.1	2.2	181	37.6	16.6	403	33.0	8.4	236	34.3	14.4
12	53	20.8	22.6	310	34.8	12.9	597	16.7	2.5	226	46.0	12.4	521	42.8	4.6	300	39.0	14.3
13	32	**9.4**	**28.1**	266	34.9	10.0	**405**	**48.1**	1.7	197	38.6	**12.2**	**420**	**36.7**	**6.9**	**216**	**36.6**	**10.2**
14	42	7.1	19.0	284	30.3	7.7	485	48.7	2.1	185	51.9	7.0	494	36.2	4.3	225	33.3	10.7
15	35	14.3	20.0	280	28.2	5.0	366	47.5	2.5	183	39.9	4.4	376	39.9	6.1	184	39.7	7.6
16	42	11.9	7.1	244	27.5	7.0	378	44.7	2.4	192	37.0	6.8	441	36.3	5.9	198	34.8	6.1
17	31	9.7	9.7	213	30.0	8.5	298	42.6	0.7	228	32.5	3.9	415	34.9	5.1	247	29.6	5.7
18	32	**18.8**	**3.1**	192	28.1	6.3	**306**	**42.5**	2.0	179	41.3	**2.2**	**305**	**33.8**	**2.6**	**198**	**32.8**	**5.6**

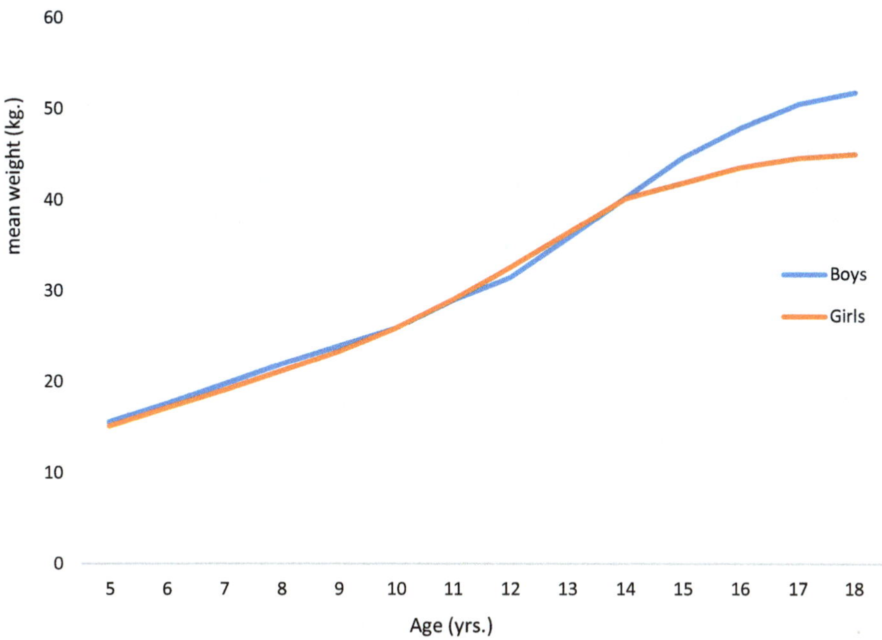

Fig. 2 Mean growth on weight of 5–18 years children in India

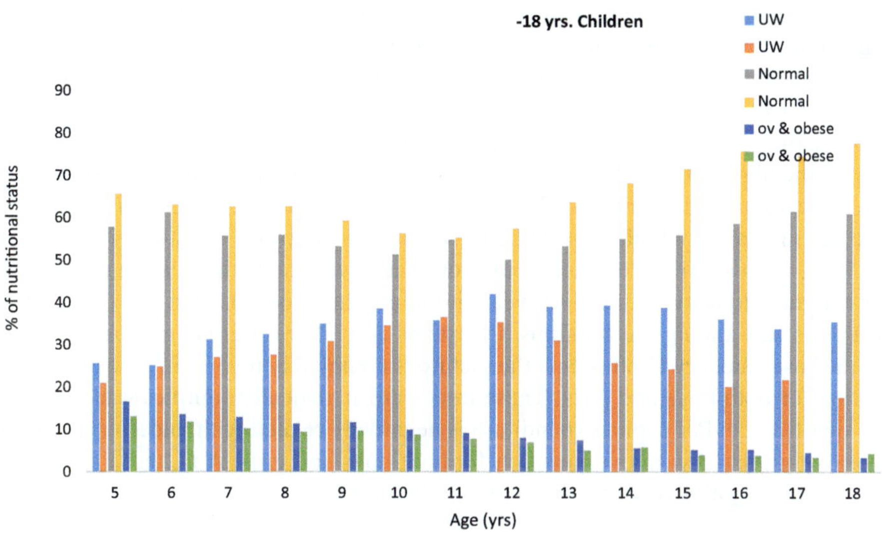

Fig. 3 Percentage of nutritional status of 5–18 yrs. children

zone is that among boys up to 10 years, overweight or obesity is 4–5 times more than underweight and then from 11 to 15 years, percentage of overweight or obese and undernutrition is more or less same and afterwards, underweight increases. In West and South zones, burden of dual malnutrition is prominent up to 12–13 years and in Northeast zones, it exists up to 7 and 10 years, respectively, but in east zone, no sign of dual burden seen because obesity is very small in percentages. This trend is thus more or less same among both boys and girls.

Table 6 represents the relationship between undernutrition or overnutrition of 5–18-year children by zone, age group, religion, per capita income separately for boys and girls. It is seen that the rate of underweight is the lowest among the boys in the Northeast zone. The highest is observed in the Central zone. In case of girls, the lowest rate is in the Northeast zone, but the highest is seen in the West zone. In case of overweight or obesity, highest obese children are found in Northeast zone, the percentages being 30.4 and 23.7 for boys and girls, respectively. It is seen that both under- and overnutrition are high among boys than girls in almost all the zones of India. Regarding relationship with age groups, underweight is bell-shaped with age and overweight or obesity is inversely related with age for both the sexes. Percentages of underweight are more among Hindus and Muslims. The Christians have least percentages of underweight children. The trend is completely opposite in case of overnutrition or obesity. The highest overweight or obese children are found among Christians. It should be noted that there are more Christians in Northeast zone than in other zones. It is also seen that per capita expenditure is positively related with overweight or obesity and is negatively related with undernutrition for both sexes. All these results are statistically significant at 5% level of significance.

Table 7 shows the relationship of socio-economic variables with under- and overnutrition of 5–18-year children through categorical logistic regression. Impacts of zones are very much prominent for both undernutrition and overnutrition because Northeast zone is taken as base. Undernutrition is very high in all other zones. The reverse picture is seen in case of overnutrition because in Northeast zone, overnutrition is the highest for both boys and girls and these results are statistically significant at 5% level of significance. Undernutrition increases and overnutrition decreases as the age increases in both boys and girls and these results are statistically significant at 5% level of significance. Religious groups are significantly associated with under- or overnutrition among the 5–18-year children in India because the risk of undernutrition is less among the Muslims, Christians and other religious groups compared to the base category Hindus, and the picture is completely reverse in case of overnutrition for both boys and girls. These trends are statistically significant at 5% level of significance. Per capita expenditure is negatively related with undernutrition and positively related with overnutrition for both boys and girls. And, these results are statistically significant at 5% level of significance.

5 Discussion

This paper presents before us a scenario of growth and nutrition status among 5–18-year children in India. The study reveals that the growth in height and weight shows

Table 5 Percentage of under- and overnutrition among 5–18-year girls in different zones of India

Age (yrs.)	Northeast			East			Central			West			North			South		
	N	UW	OW/Obese	N	UW	OW/Obes	N	UW	OW/Obese	N	UW	OW/Obese	N	UW	OW/Obese	N	UW	OW/Obese
5	25	**16.0**	52.0	243	21.4	14.4	413	20.6	6.3	133	22.6	23.3	347	21.0	11.0	165	21.2	18.8
6	32	15.6	46.9	252	20.2	15.9	397	25.4	4.8	119	31.9	16.8	369	25.7	8.4	166	25.3	20.5
7	32	18.8	31.3	321	24.9	16.8	467	29.8	3.2	189	29.1	18.5	411	25.8	5.8	184	26.1	14.7
8	30	3.3	43.3	243	23.5	13.6	459	33.1	3.7	143	34.3	12.6	336	25.6	3.9	188	22.9	20.7
9	37	10.8	40.6	213	30.5	9.9	347	34.3	2.6	166	35.5	16.9	347	32.3	7.2	191	22.5	15.7
10	56	7.1	35.7	337	28.2	10.4	525	45.1	2.3	180	36.1	11.1	411	33.8	6.6	213	27.2	18.3
11	43	14.0	18.6	211	36.0	10.4	335	46.0	2.4	158	34.8	12.7	326	35.3	4.6	204	30.9	13.7
12	61	13.1	26.2	349	31.2	10.0	568	39.4	3.7	185	38.9	9.2	461	36.0	5.0	233	33.9	8.6
13	44	**13.6**	**13.6**	272	22.1	5.9	**436**	**36.7**	2.3	204	37.7	**5.4**	**386**	**29.5**	**3.6**	**238**	**31.5**	**10.5**
14	34	11.8	11.8	308	20.5	5.8	480	27.1	4.0	220	36.4	9.5	417	24.7	4.1	229	24.0	9.2
15	48	4.2	4.2	276	19.6	5.1	423	27.0	1.9	187	23.5	6.4	382	27.5	2.9	233	24.5	6.9
16	38	0.0	7.9	287	13.6	3.1	114	23.8	2.7	192	25.0	2.6	396	21.5	3.5	245	18.8	8.2
17	31	9.7	3.2	232	15.9	3.9	387	22.0	2.6	203	28.6	3.4	437	19.7	3.9	245	26.9	3.7
18	21	**4.8**	**0.0**	255	12.5	5.5	**390**	**17.7**	3.3	191	21.2	**4.2**	**387**	**19.9**	**3.9**	**257**	**14.4**	**6.6**

Table 6 Relationship between socio-economic variables and 5–18-year children's undernutrition and overnutrition in India

Dependent variables	Boys				Girls			
	N	UW	OW or obese	Chi-square	N	UW	OW or obese	Chi-square
Zone				845.63 df= 10				658.39 df= 10
Northeast	536	15.9	30.4		532	10.2	23.7	
East	3744	30.1	11.4		3799	22.9	9.3	
Central	5979	40.6	3.4		6068	30.9	3.3	
West	2621	39.2	12.9		2470	31.6	10.2	
North	5927	34.1	7.4		5413	27.0	5.2	
South	2985	32.7	14.1		2991	25.0	11.9	
Age groups (yrs.)				398.331 df=4				455.85 df= 4
5–9	7690	29.9	13.3		6965	26.3	10.9	
10–15	9963	39.2	7.7		9673	31.3	6.5	
16–18	4139	35.3	4.6		4635	19.9	3.9	
Religion				103.04 df= 6				91.107 df= 6
Hindu	17435	36.3	8.5		16888	27.7	7.0	
Muslim	3169	32.9	10.9		3335	26.7	8.2	
Christian	379	21.9	17.2		370	17.3	18.4	
Others	809	26.5	10.6		680	22.4	7.5	
Per capita expenditure				212.247 df=10				230.428 df=10
<10,000	4102	38.4	7.8		4663	30.7	5.6	
>10000–<14000	4521	37.9	7.8		4616	29.1	5.4	
>14000–<19000	4313	38.0	7.5		4198	28.4	6.7	
>19000–<27000	4055	34.3	9.5		3704	25.7	8.8	
>27000–<50000	3567	29.2	11.4		3059	21.9	10.6	
>50000	1052	25.5	15.6		859	18.5	13.0	

All the tests are significant at 5% level

a steady increase from 5 to 18 years. Juvenile growth spurt in height among boys occurs at the age of 6–8 years while among girls, it occurs between 5 and 6. But in case of weight, there is no prominent sign of juvenile growth spurt among both boys and girls. Among boys, adolescent growth spurt in height occurs at 12 years and continues up to 15 years and among girls, it starts around 10 years and continues up to 14 years. In case of weight, among boys, and girls, adolescent growth occurs from age 12 to 15 years and 10 to 14 years respectively.

From different grades of nutritional status, among boys and girls, it is seen that percentage of undernutrition increases sharply up to 12 years and then it decreases with very marginal effect. Among girls, the magnitude of decrement of undernutrition is very sharp. In case of overweight or obesity, it is very well marked up to the age of 10 and then it decreases sharply and percentage of obesity resides between 3 and 5% at the age of 18. So it may be said that percentage of overweight or obesity is very much alarming among the children of pre-juvenile to pre-adolescent period

Table 7 Categorical logistic regression between socio-economic variables and 5–18-year children's undernutrition and overnutrition in India

Dependent variables	Boys		Girls	
	Underweight	Overweight	Underweight	Overweight
Zone				
Northeast	1.00	1.00	1.00	1.00
East	1.899*	0.310*	2.264*	0.431*
Central	3.056*	0.082*	3.484*	0.133*
West	3.167*	0.351*	3.983*	0.421*
North	2.674*	0.169*	3.276*	0.180*
South	2.466*	0.375*	3.017*	0.447*
Age groups (yrs.)				
5–9	1.00	1.00	1.00	1.00
10–15	1.556*	0.503*	1.323*	0.506*
16–18	1.370*	0.256*	0.733*	0.281*
Religion				
Hindu	1.00	1.00	1.00	1.00
Muslim	0.893*	1.239*	0.995	1.227*
Christian	0.684*	0.976	0.768	1.523*
Others	0.685*	1.102	0.824*	1.027
Per capita expenditure				
<10,000	1.00	1.00	1.00	1.00
>10000–<14000	0.960*	0.961	0.915*	0.956
>14000–<19000	0.925*	0.940	0.870*	1.252*
>19000–<27000	0.777*	1.213*	0.767*	1.707*
>27000–<50000	0.626*	1.497*	0.630*	2.127*
>50000	0.513*	2.299*	0.521*	2.743*

®=Reference category; *Significant at 5% level

such as from 5 to 11 years in India. And the result confirms the marked existence of double burden of malnutrition in this period. From zone wise comparison, it is seen that in the Northeast zone, the lowest percentage of undernutrition is as high as 30.4% and the highest percentage of overweight or obese children is 23.7%. Very low percentages of overweight or obese children are found in Central zone then followed by East zone. This trend is same for both the boys and girls. It is seen that both under- and overnutrition are high among boys than girls in all the zones of India. Regarding dual burden of malnutrition, it is seen that in Northeast zone, up to 10 years, overweight or obesity is 4–5 times more than underweight and then from 11 to 15 years, the percentage of overweight or obese and undernutrition is more or less same and afterwards, underweight increases. In West and South zones, burden of dual malnutrition is prominent up to 12–13 years and in North and East zones, it exists up to 7 and 10 years, respectively, but in east zone, no sign of dual burden exists because obesity is very less in all the age groups. This trend is more or less same among both boys and girls.

So far as religion is concerned, the Hindus are seen to have the maximum percentage of underweight children. The least is among the Christian and other religious groups for both the sexes, but the trend is completely opposite in case of overnutrition or obesity. It is also seen that per capita expenditure is positively related with overweight or obesity and is negatively related with undernutrition for both sexes. Unlike the studies, which were mostly carried out on pre-school and post-school children taking a short age interval, this study describes nutrition status of children over a long range of ages covering both juvenile and adolescent growth spurts. And this study also shows the existence of dual burden of malnutrition in Juvenile to pre-adolescent Indian children. This type of studies is rare, because it not only includes the picture of both juvenile and adolescent growth spurt but also the edge of adulthood in India. The existence and extent of double burden of malnutrition are also discussed in this paper.

Acknowledgements We are thankful to both the anonymous referees for their valuable comments and suggestions, which enabled us to improve the paper substantially to the present version.

References

Barry, M. P., & Penny, G.-L. (2004). The nutrition transition: worldwide obesity dynamics and their determinants. *International Journal of Obesity*, *28*, Supplement 3: S2–S9.

Borooah, V. K. (2005). The height-for-age of Indian children. *Economics and Human Biology, 3,* 45–65.

Caballero, B. (2005). A nutrition paradox—underweight and obesity in developing countries. *The New England Journal of Medicine, 352,* 1514–1516.

Chhatwal, J., Verma, M., & Riar, S. K. (2004). Obesity among pre-adolescent and adolescents of a developing country (India). *Asia Pacific Journal of Clinical Nutrition, 13,* 231–235.

Coly, A. N., Milet, J., Diallo, A., Ndiaye, T., Bénéfice, E., Simondon, F., et al. (2006). Preschool stunting, adolescent migration, catchup growth, and adult height in young Senegalese men and women of rural origin. *Journal of Nutrition, 136,* 2412–2420.

Dietz, W. H. (2001). Overweight and precursors of type 2 diabetes mellitus in children and adolescents. *The Journal of Pediatrics, 138,* 453–454.

Doak, C. M., Adair, L. S., Bentley, M., Monteiro, C., & Popkin, B. M. (2005). The dual burden household and the nutrition transition paradox. *International Journal of Obesity, 29,* 129–136.

Doak, C. M., Adair, L. S., Monteiro, C., & Popkin, B. M. (2000). Overweight and underweight Coexist within Households in Brazil, China, and Russia. *The Journal of Nutrition, 130,* 2965–2971.

FAO (2006). *The state of food insecurity in the World 2000: Food insecurity—when people live with hunger and fear starvation*. Rome: FAO. Retrieve January 12, 2006 from www.fao.org.

Gorstein, J., Sullivan, K., Yip, R., de Onis, M., Trowbridge, F., Fajans, P., & Clugston, G. (1994). *Bulletin of the World Health Organization*, *72*, 273–283.

Indian Human Development Survey-II (IHDS-II), (2011–12). Inter-university Consortium for political and social research. P.O. Box 1248. Ann Arbor, Michigan 48106. Retrieved from www.icpsr.Umich.edu.

Jelliffe, D. B. (1966). The assessment of nutritional status of a community. *WHO Monog series*. (No. 53, pp. 1–271).

Kennedy, G., Nantel, G., & Shetty, P. (2006). Assessment of the double burden of malnutrition in six case study countries. FAO, the double burden of malnutrition: case studies from six developing countries. In *Food and Nutrition Paper 84. Rome: Food and Agricultural Organization of the United Nations* (pp. 1–41).

Khadilkar, V. V., & Khadilkar, A. V. (2004). Prevalence of obesity in affluent school boys in Pune. *Indian Pediatrics, 41,* 857–858.

Laxmaiah, A., Nagalla, B., Vijayaraghavan, K., & Nair, M. (2007). Factors affecting prevalence of overweight among 12 to 17 year old urban adolescents in Hyderabad, India. *Obesity (Silver Spring), 15,* 1384–1390.

Mak, K. K., & Tan, S. H. (2012). Underweight problems in Asian children and adolescents. *European Journal of Pediatrics, 171,* 779–785.

Martorell, R., & Ho, T. J. (1984). Malnutrition, morbidity and mortality. In W. H. Mosley & L. C. Chen (Eds.), *Child survival: Stratigies for research. Population and development review; a supplement to* (Vol. 10. pp. 49–69). Cambridge: Cambridge University Press.

Panjikkaran, S. T., & Kumari, K. (2009). Augmenting BMI and Waist-Height Ratio for establishing more efficient obesity percentiles among school children. *Indian Journal of Community Medicine, 34,* 135–139.

Popkin, B. M. (1993). Nutritional patterns and transitions. *Population and Development Review, 19,* 138–157.

Raj, M., Sundaram, K. R., Paul, M., Deepa, A. S., & Kumar, R. K. (2007). Obesity in Indian children: Time trends and relationship with hypertension. *The National Medical Journal of India, 20,* 288–293.

Rao, N. F., Camnath, T., & Sastry, J. G. (1986). Diet and nutrition in urban areas. *The Proceedings of the Nutrition Society, 32,* 91–99.

Reilly, J. J., Methven, E., McDowell, Z. C., Hacking, B., Alexander, D., Stewart, L., et al. (2003). Health consequences of obesity. *Archives of Disease in Childhood, 88,* 748–752.

Subramanyam, V., Jayashree, R., & Rafi, M. (2003). Prevalence of overweight and obesity in affluent adolescent girls in Chennai in 1981 and 1998. *Indian Pediatrics, 40,* 332–336.

World Health Organization Multicentre Growth Reference Study Group. (2006). *WHO Child growth standards: length/ height for age, weight for age, weight for length, weight for height and body mass index for age methods and development*. Geneva: WHO.

Retrospective Longitudinal Studies on Lifestyle and Social Environment in North-Eastern Tribes

Ratan Dasgupta

Abstract Retrospective longitudinal studies are conducted on lifestyle and social environment on a sample of 270 tribal individuals from North-Eastern tribes in Tripura. The study is of relevance to the welfare of children, involving their families and the local community. Time of the status enquiry covers a span of more than 5 years, viz. 20 September 2011–30 January 2017. Lifestyle status of villagers, consisting of 'food, shelter, and health' at present time, 2 years back and 5 years back from the time of interview are enquired, and the status is categorised as comfortable and not comfortable. Percentage of positive response recorded indicates a general improvement of lifestyle status over time. Question on social environment in individuals' locality of residence is asked for the present time and retrospective time of 5 years back from the date of the interview to assess the state of improvement over years. The scores assigned are peaceful (score 3), moderately peaceful (score 2), disturbed (score 1) and highly disturbed (score 0), and the scores are then added up and expressed as percentage with respect to maximum possible score at a particular year. This provides a picture of (partial) fulfilment of individuals' expectation on social environment with respect to the highest possible state. Growth curves of lifestyle and social environment over time indicate a general improvement of tribal status over years.

Keywords Retrospective longitudinal study · Lifestyle status · Proliferation rate

Classifications 62P25, 65D10

R. Dasgupta (✉)
Theoretical Statistics and Mathematics Unit, Indian Statistical Institute,
203 B.T. Road, Kolkata 700108, India
e-mail: rdgupta@isical.ac.in; ratandasgupta@gmail.com

R. Dasgupta (ed.), *Advances in Growth Curve and Structural Equation Modeling*,
https://doi.org/10.1007/978-981-13-0980-9_9

1 Introduction and the Methodology

Retrospective longitudinal data are of interest when comparing the present situation with past to see the quantum of improvement, if any; in the way of looking back. Retrospective analysis may sometime leads to confounding when important covariates are not recorded.

When existing data such as medical records or claims database are available, then retrospective longitudinal studies may be made possible with advantage, e.g. see Cabrera et al. (2015). Peaceful environment with comfortable lifestyle is of importance in development of children that involves their family and the community as well, e.g. see McAuley et al. (2006). Like the procedure followed in longitudinal studies reported in Dasgupta (2018), the individuals in the retrospective longitudinal studies reported here are interviewed at different places of congress like students in different schools, individuals in place of work, marketplaces, church and villages during data collection in interiors of Tripura, following the procedure of stratified sampling.

At start of the investigation, the questions were asked and explained verbally by the interviewer. Some answers were provided by tribal individuals in writing, and others were recorded by interviewer. At the later stage of investigation, written questionnaires were provided and the answers were given by the individuals in between the blank spaces of printed questions.

The retrospective longitudinal study on lifestyle of tribal individuals in Tripura reported here covers a time span of more than 5 years on the effected background of unrest from rebel insurgency.

Individuals were interviewed on status of local environment in terms of the scores are peaceful (score 3), moderately peaceful (score 2), disturbed (score 1) and highly disturbed (score 0). Twenty persons are interviewed in the year 2017 and 187 in the year 2016.

Main results are stated in Sect. 2, and the status of tribal lifestyle is shown in figures. Results are further discussed in Sect. 3.

2 Main Results

Growth curve of percentage of individuals reporting comfortable lifestyle increases over years. Lowess regression indicates a gradual improvement of tribal lifestyle status over time. Starting from 11.49%, in Fig. 1, the number rises to 70.10%; during the time span 2011–2017. A similar improvement is seen for local environment status. Percentage score based on different state on 'peace in locality' in Fig. 2 goes up from 73 to 94% in the same time span in the background of rebel insurgency. Thus, improvement is observed in terms of the criterion, 'tribal lifestyle status' and 'peace in locality'. The results are explained further in figures and elaborated with legends (Figs. 3, 4 and 5).

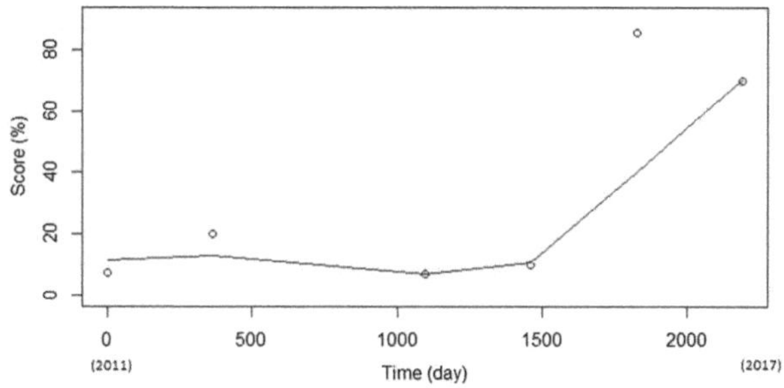

Fig. 1 Percentage of individuals reporting comfortable lifestyle. The query on comfortable lifestyle status is asked to ($187 + 20 = 207$) tribal individuals at the year of interview, i.e. status at 'present' time point (2016, 2017), along with on status at 2 years back (2014, 2015), and 5 years back (2011, 2012), thus generating 6 times points in all; this provides a set-up of retrospective longitudinal study. The binary score 1 and 0 are assigned for comfortable, and not comfortable lifestyle, respectively. Percentage of tribal individuals reporting 1 at a particular time is lowess regressed with $f = 2/3$ to obtain this figure. The growth curve indicates a gradual improvement of tribal lifestyle status over time

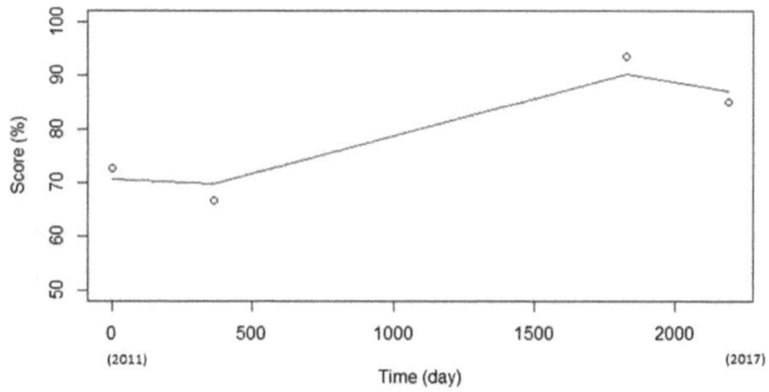

Fig. 2 Score of local environment in 5 years. In this figure, we show status of local environment in terms of the scores as peaceful (score 3), moderately peaceful (score 2), disturbed (score 1) and highly disturbed (score 0). Twenty persons are interviewed in the year 2017 and 187 in 2016. Total number of individuals interviewed is 207. Out of 187 individuals interviewed in the year 2016, only 64 person gave score 3 to the local environment that existed 5 years back, i.e. in 2011, only 101 persons gave score 2, only 14 persons gave score 1, and the remaining 8 persons gave score 0 to local environment of the year 2011. Maximum possible score per person is 3. Thus, the percentage score of the year 2011 is $100(3 \times 64 + 2 \times 101 + 1 \times 14 + 0 \times 8)/(3 \times 187) = 72.73$, with respect to the maximum possible score (3×187). In a similar manner, score of the year 2012 is 66.67; for the years 2016 and 2017, the scores are 93.58 and 85, respectively. The scores shown in the figure indicate improvement of local environment over time. The smooth curve is drawn by smooth.spline in R with df = 3.3

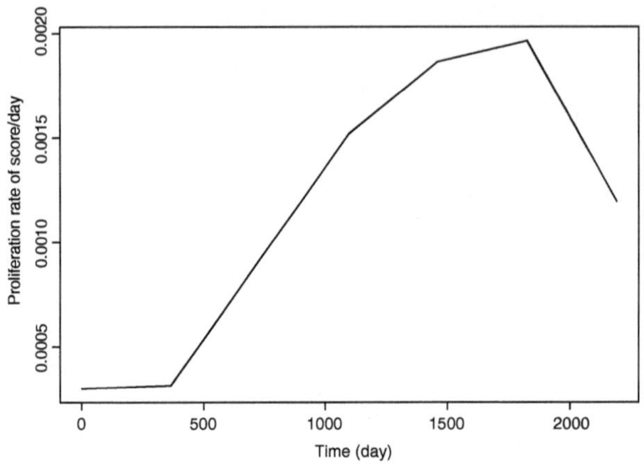

Fig. 3 Proliferation rate of 207 persons lifestyle: wt. exp(−1.9 x); spline. Proliferation rate $\frac{d}{dt}\log y = \frac{1}{y}\frac{dy}{dt}$ is a scaled version of velocity $\frac{dy}{dt}$. This measure is independent of the choice of unit used in measuring y. Computation is based on a technique proposed in Dasgupta (2015), with exponentially decaying weights attached to empirical slopes computed from data pairs at different time points with respect to a fixed time point t of interest. More weights are given to data points near the time t of derivative computation, and less weights to distant time points from t. Here, we consider a sharply decreasing weight function $w(x) = \exp(-1.9\ x)$ that assigns more weights to observations near time point of calculating derivative. Weighted mean of these empirical slopes at derivative stage and smooth.spline with spar = 0.0001 at smoothing stage in SPlus provide proliferation rate at time point t. The choice of weight function $w(t) = \exp(-1.9\ x)$ results in the proliferation curve that is in conformity with the pattern of growth curve. The curve starting from the year 2011 has a high peak at 1827 day. The curve then drops towards the baseline at zero over time in this figure

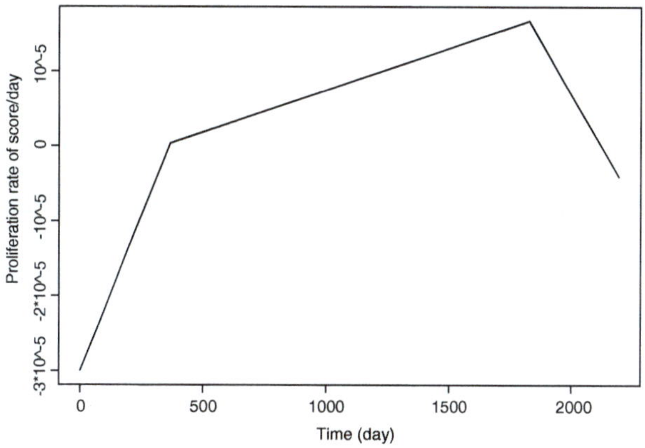

Fig. 4 Proliferation rate of local environment score: wt. exp(−0.00131 x); spline. We compute the proliferation rate based on the growth curve shown in Fig. 2. For local environment score y with growth curve computed in Fig. 2 by smooth.spline produced the values (70.04, 68.25, 88.88, 87.86) at time points (0, 366, 1827, 2192) respectively. The proliferation rate curve of lifestyle status with spar = 0.0001 for above values is shown in this figure. The rate goes up with a steep start, then achieves a peak slowly, and then goes down in a steep manner towards zero. The choice of weight function $w(t) = \exp(-0.00131\ x)$ results in the proliferation curve that is in conformity with the pattern of growth curve

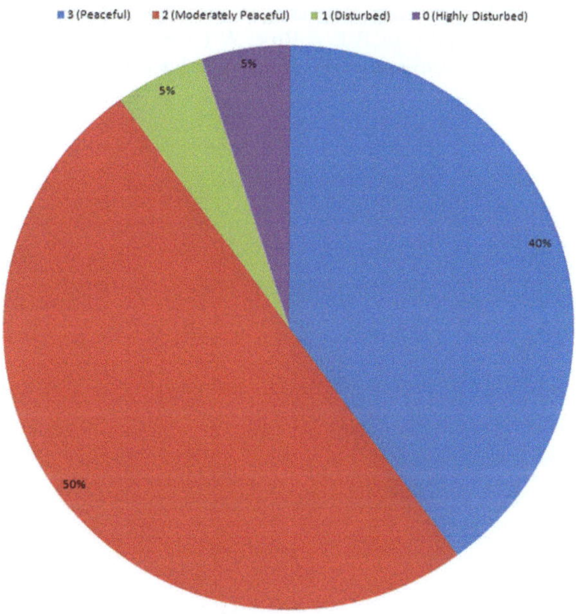

Fig. 5 Pie Diagram of local environment in general during 2011–2017. This figure shows the pie chart based on the percentage of tribal individuals who considered the time zone 2011–17 to be peaceful in general (40%), moderately peaceful (50%), disturbed (5%), and highly disturbed (5%). A large percentage (90%) of persons feels that the environment of locality at home is moderately peaceful and better

3 Discussion

The nonparametric growth curves indicate a gradual improvement of tribal lifestyle status over time. The percentage of tribal individuals who considered the time zone 2011–17 to be peaceful in general are high, 40%; those who considered it to be moderately peaceful is 50%. The percentage of persons considered the environment to be disturbed is 5%; and the remaining individuals, those considered the environment to be highly disturbed are of 5% of the population. A large percentage (90%) of persons feels the environment of locality at home is moderately peaceful and above.

References

Cabrera, C., Brunelli, S. M., Rosenbaum, D., Anum, E., Ramakrishnan, K., Jensen, D. E., et al. (2015). A retrospective, longitudinal study estimating the association between interdialytic weight gain and cardiovascular events and death in hemodialysis patients. *BMC Nephrology, 16*, 113. https://doi.org/10.1186/s12882-015-0110-9.

Dasgupta, R. (2015). Rates of convergence in CLT for two sample u-statistics in non iid case and multiphasic growth curve. In R. Dasgupta (Ed.), *Growth curve and structural equation modeling*. Springer Proceedings in Mathematics & Statistics (Vol. 132, pp. 35–58).

Dasgupta, R. (2018). Longitudinal studies on mathematical aptitude and intelligence quotient in North Eastern tribes. Appearing in this volume as chapter 1.

McAuley, C., Pecora, P. J., & Rose, W. (2006). *Enhancing the well-being of children and families through effective intervention*. London: Jessica Kingsley Publishers.

Microevolution Among the *Yanadi* Tribe: Population Structure and Sex Differences in Anthropometric and Dermatoglyphic Characters

T. S. Vasulu

Abstract Microevolutionary trends in anthropometric and quantitative dermatoglyphic traits have been investigated among the five regional populations of the Yanadi tribe, who live in different regions and differ in their life style and population structural variables. Of the five, CY and IY live as hunters-gatherers in coastal areas where as rest three P1, HF and P2 survive as agricultural labourers and agricultural landholders in hill-forest (HF) and plateau regions (P1, P2) in three districts in Andhra Pradesh. The regional differences in some of the population structure variables; viz., marriage pattern, inbreeding etc., observed among the regional populations indicate the possibility of microevolutionary trends. This has been investigated in case of morphology and dermatoglyphics traits. The results show significant differences both in size and shape components of morphology and in some of the quantitative characters e.g., triradial numbers, finger ridge count and in pattern combination indices among the regional populations. The CY and IY coastal populations differ from the three inland populations of pre-agricultural and agricultural populations. The results show that males and females within a regional population are more similar in their morphology and dermatoglyphic traits than when compared with their counter parts in other regional populations. The dendrograms show consistency in the clustering pattern and also show differences between males and females, which is in agreement with the expectation based on population structure variables. The pattern combination analysis (PCA) shows significant differences in some of the pattern combination types of monmorphic and trimorphic pattern types (PCT-tr) and significant sex differences in quantitative PCA method where females show the absence of tertramorphic pattern ratios.

Keywords Breeding population · Inter and intra population differences · Size and shape component · Male-female differences · Correlographs · Anthropometric and dermatoglyphic profiles · Clustering · Finger pattern combination index (FPCI)

T. S. Vasulu (✉)
Indian Statistical Institute, 203 B.T. Road, Kolkata 700108, India
e-mail: vasulu@gmail.com

R. Dasgupta (ed.), *Advances in Growth Curve and Structural Equation Modeling*,
https://doi.org/10.1007/978-981-13-0980-9_10

1 Introduction

The study of microevolution among human populations involves at least three major issues: (a) the identification and type of breeding population suitable for the study, (b) selection of traits or characters (phenotypic traits, genetic, molecular genetic and polygenetic traits etc.,) and methodological issues concerning the identification of breeding populations, and (c) appropriate statistical techniques and theoretical populations genetics models to test specific hypotheses related to population structure variables and type of characters selected for the study and nature of type of breeding populations.

A special focus for the microevolutionary studies are populations who are relatively undisturbed, isolated or insular and tribal populations, since their way of life is similar or comparable to the situation during the early periods of human existence (Neel 1967, 1970). In particular, some of the tribal populations living in traditional way of life or in transitional stage are influenced by the environmental conditions that foster quick changes in demographic and genetic structure. This gives us a rare advantage to investigate the biological consequences and microevolutionary trends (Salzano 1978). These populations are also important since a majority of these populations are getting disappeared or getting merged with the civilized world at large.

As far the selection of traits for the investigation of genetic differentiation and microevolution is concerned, classical genetic traits and molecular genetic markers are best suited, however, poly genetic traits, such as anthropometric and dermatoglyphic traits which supposedly encompass a major part of the genome are also of importance. In this respect these polygenetic characters or traits e.g., anthropometric traits etc., are influenced by both genetic and non-genetic factors, especially, population structure variables (which vary widely among populations). Whereas dermatoglyphic traits are mostly governed by genetic factors, pose an interesting area of study. Recent developments in human genome, especially genome-wide scan and search for genetic clues etc. has opened up an opportunity to investigate the microevolutionary studies of complex phenotypes such as height, BMI, body size and shape etc. (e.g., Lettre et al. 2008; Nettle 2002a, b; Sanna et al. 2008; Weedon et al. 2008; Soranzo 2009; Lango Allen et al. 2010; Yang et al. 2010; Amato et al. 2011).

2 Studies on Mircroevolution

2.1 Microevolution and Population Structure

The initial studies among Amerindian tribes (e.g., Xavante, Markiritare, Yanomama etc.) by Neel and his co-workers have revealed that the population structure of these tribes is governed by fission-fusion dynamics. In simple tribal societies, fission or division can happen due to internal conflict or fights between different sections or say e.g. clans and fusion or merger between clans or factions of the tribe can also happen e.g., in situations where they cooperate together to face natural calamities or adversaries. In genetic point of view, fission and fusion population structure among a tribal society can lead to non-random sampling of gene pools and it can have a remarkable effect on the genetic diversity of the population (Neel and Salzano 1967; Neel 1970; 1978; Neel and Ward 1970; Neel et al. 1974). The possible consequences of fission-fusion population structure on the changes in genetic diversity have been examined in case of genetic, dermatoglyphic and anthropometric characters among the Amerindian tribes (Spielman 1973a, b; Neel and Ward 1970; Rathhammar et al. 1973, 1977). Similar such studies conducted among other world populations substantiate the findings of the Amerindian tribes. A number of studies have reported the effect or influence of population structure variables on the genetic differentiation (Hiernaux 1966; Neel and Salzano 1964a; b; 1967; Cavalli Sforza 1969; 1974; Linvingstone 1969; 1972; Erikson et al. 1970; Workman and Niswander 1970; Workman et al. 1975; Friedlander 1975; Chagnon et al. 1970; Chagnon 1972; Glas et al. 1952; Jorde 1980; Kirk 1982; Kraus and White 1956; Kuchemann and Lasker 1980; Spielman et al. 1973; Friedlander 1975; Lees and Crawford 1976). Oliver and Howells (1956, 1957) in two classic papers demonstrated that microevolutionary changes leading to genetic differentiation are not limited to single gene traits but they are also reflected in the variation of continuous (polygenic) anthropometric variations, in spite of their modifications by environmental factors (Lasker 1946, 1960; Leslie 1980, 1985; Marquer and Jakobi 1976; Nettle 2002a, b; Martin 1970). Some studies have shown that quantitative traits are also subject to rapid selection although the effect of random drift may be less readily discernible in these multifactorial traits (Ganguly and Pal 1972; Demarchi and Marcellino 1998; Fleishman 1980; Harrison and Boyce 1972; Loesch and Liseiwicz 1976; Mielke and Crawford 1973; Mukherjee 1989).

Several studies have investigated the influence of population structure variables e.g. marriage, migration, on quantitative traits, e.g., anthropometric variation among tribal, regional, within ethnic and breeding populations (Robers 1956; Basu 1969; Mahalonobis et al. 1941; Sanghvi 1953; Majumdar and Rao 1960; Katayama and Toyomasu 1979; Karve and Malhotra 1968; Malhotra 1974; Malhotra et al. 1978; Martin 1970; Das 1981, 1985; Mukherjee and Chakravartii 1964; Papiha 1983). In general, microevolutionary studies have accounted changes in the degree and pattern of differentiation as a result of cultural factors and history, migration etc. (Livingstone 1969; Workman and Niswander 1970; Friendlander et al. 1971; Spielman 1973a; Friedlander 1975; Lees and Crawford 1976). A few studies have considered intra- and-inter population variation with respect to a set of anthropometric characters and as well overall morphological variation with respect to endogamy, isolation etc. (Mahalonabis et al. 1941; Sanghvi 1953; Majumdar and Rao 1960; Karve and Malhotra 1968; Das 1985). Studies on microevolution also have considered the influencing factors related to cultural aspects, such as: mating pattern (Oliver and Howels 1956, 1957; Hiernaux 1966; Das 1981). A few studies also have considered intra- and-inter population variation in anthropometric characters and morphological variation among tribes and castes with respect to factors like: migration, fusion-fission etc. (e.g., Spielman et al. 1973; Lees and Crawford 1976; Smouse et al. 1981; Lees and Relethford 1982; Blangero 1990; William-Blangero and Blangero 1990; Vasulu 1993).

Further, Chakraborty and Malhotra (1981) have shown that the rate of differentiation in dermatoglyphic character under the assumption of random drift is not proportional to the time of divergence. It is somewhat slower than that indicated by genetic markers (Wright 1965; Nei and Chakrobarti 1977). Random drift among population also gets influenced by migration as well (Rogers 1968). The role of natural selection as possible factor among the primitive populations of Amerindian tribes has been considered by Salazano (1975). They however argue that observed levels of variation of number of triradii on fingers in different populations are largely the result of random drift. In a study of Taiwan populations, the dermatoglyphic characters reflected ethnohistorical similarities between related tribes to a great extent, whereas anthropometric traits displayed some effect of microevolutionary changes (Chai 1972) as suggested by Howels (1966). Some studies among Indian populations also show slower microevolutionary changes through generations in dermatoglyphic traits between related populations than in anthropometric traits (Roberts and Coope 1972; Penrose 1965; Mukherjee 1978, 1989; Wang and Barrantes 2009).

2.2 *Microevolution and Sex Differences*

A few studies have also shown a greater within village homogeneity in some of the biological traits in relation to population structure variables, especially in societies under the patrilineal, patrilocal and patriarchal system; where the village endogamy is characterized by marital migration of brides from other villages whereas brothers live in the same village. In genetic sense it leads to greater within morphological homogeneity among males than their wives. For example, studies among Amerindians have observed a greater within village morphological homogeneity between sexes when compared to other villages (Spielman 1973a, b; Neel and Ward 1973). Lees and Relethford (1982) in a study among rural western Ireland have observed greater anthropometric variation among female migrants than male migrants (Relethford et al. 1980). Similar studies conducted among Jirles of Nepal by William-Blangero and Blangero (1990) have found inter-village sex differences, especially, greater variance in cranial measurements and other anthropometric characters among females than males. Other few studies include Niswander et al. (1970) among Papago Indians, among Australian aborigines of Arnhem Island (White and Parson 1973) and among Tunisian population (Namouchi 2011), among Albanian of Kosovo (Temaj et al. 2010, 2012) and a recent study among pre-Hispanic communities in Andes (Rodriguez Florez and Colantonio 2015). In India similar findings have been reported among caste-clusters of Brahmin and Vaisya communities by Rao et al. (1983) and in anthropometric and size and shape components of morphology among the Yanadi tribal populations (Vasulu 1991, 1998, 1999).

In continuation with the above study, present investigation reports microevolutionary trends with respect to anthropometric and quantitative dermatoglyphic characters among the Yanadi tribe of Andhra Pradesh. The tribal population Yanadi, provide many features that are suitable to investigate microevolution. These include, e.g., wide geographical distribution as a result of migration resulting isolation and formation of regional breeding populations for generations in different ecological zones, exhibit wide differences in subsistence pattern, common ethno-historical origin and other associated cultural traits e.g., patriarchal system, practice of consanguineous marriage and endogamy etc. These provide an opportunity to investigate the process of genetic differentiation in relation to changes in population structure, subsistence pattern and other cultural differences that promote microevolution among regional populations of the tribe.

3 Materials and Methods

3.1 Population: The Yanadi

The *Yanadi*, is the second largest tribal population—after *Chenchu*—of Andhra Pradesh. Though they are mostly concentrated in southern districts, a majority of them inhabit in three southern districts of Nellore, Chittoor and Cuddapah (Stuart 1891; Thurston and Rangachari 1901; Thurston 1909; Census 1971). They were supposedly the earliest aboriginal tribes that settled in the coastal and plateau regions. The antiquity and origin of the Yanadi is difficult to trace and the earliest information about the presence of the tribe can be traced as early as from the period of Vijayanagara Kings (1600 AD). The ethnographic information indicates that the Yanadi tribe were migrating from their original abode to other nearby regions, primarily as a result of livelihood and survival reasons.

The word Yanadi the name given to them possibly refers to a word in Sanskrit—"*Anadi*" denoting one who is ancient or early inhabitants or aboriginals with unknown antiquity. Aiyappan, a leading pre-independent Anthropologist is of the opinion that the term Yanadi is possibly derived from the word "*Einar*" which in old Tamil (Thurston 1909; Iyer 1968) refers to inhabitants of the Palai or waterless region. However, historical documents reveal that the Yanadi are related (ethnically and by antiquity) to the *Chenchu* tribe of Nallamala or Nallamalai hills, the largest tribal population of A.P. (Reddy 1945; Iyer 1968; Ragahvaiah 1962; Reddy et al. 1982a). There are two recognized endogamous groups of the Yanadi exist: 1. The "*Challa*" and 2. The "*Manchi*". The Challa Yanadi, are primarily live as nomads and feed on refuge materials and are considered "inferior". They neither inter-dine, intermarry nor intermix freely with the other groups (Raghavaiah 1962). There are no available written records about the distribution of the Challa Yanadi and the survey suggests that they are numerically small (possibly a few thousands) and are restricted to some parts of Nellore district (A.P.).

In Andhra Pradesh they are called as *Yanadi* and speak local language *Telugu* whereas in adjacent districts of Tamil Nadu they are referred as *Irula* and speak the local language *Tamil*, while the border districts of Andhra Pradesh and Tamil Nadu the tribe speak both the languages as this has been the natural process of adaptive mode of population settlement of tribes and castes in India (Rakshit 1972). Interestingly in North-western parts of Nilagiri Hills, bordering Karnataka and Kerala states there are tribal populations called by the same or similar name *Irular* as well, however, they are morphologically different from the eastern Yanadi or Irula (Aiyappan 1948; Census 1961, 1971; Iyer 1968). There are also no available written records or information about their original language spoken during their historical settlement in southern parts (Iyer 1968; Raghavaiah 1962). There are also hardly any genetic or biological studies among the Yanadi, except blood group and serological or biochemical studies (e.g., Reddy et al. 1982b).

They inhabit over different ecological regions: coastal, island, plateau and forest hill regions and differ in their subsistence economy and other associated cultural attributes. For example, on the islands of Pulikate lake, the tribe live as hunters-and-gatherers/incipient-agricultural stage; in plateau regions, they are mostly composed of casual and agricultural labourers, except a few agriculturists, and in hilly-forest regions they survive as gatherers and occasional agricultural labourers (except one or two agriculturalists. As a result the Yanadi population has undergone cultural transformation and change in the population structure, especially local endogamy around nearby settlements thus over time (generations) has possibly lead to local regional breeding populations, thus paving the way or opportunity for potential microevolutionary changes.

3.2 The Present Study

A field work has been conducted during 1977–1980 in 13 settlements distributed in 3 islands of Pulikate lake surrounding Sriharikota and also in plateau and hill forest regions in Nellore, Chittoor and Cuddapah districts of Andhra Pradesh (AP) with an objective of identifying the local regional breeding populations, endogamy, admixture rate, detailed demography and investigate the possible microevolutionary trends or genetic differentiation in anthropometric and dermatoglyphic features as a result of isolation and formation of breeding populations with differences in associated cultural differences. The location of the 13 settlement studied for the study is shown in Fig. 1. A more detailed description of the detailed field work, method of selection of locations, villages and identification of breeding populations have been described earlier (Vasulu 1989).

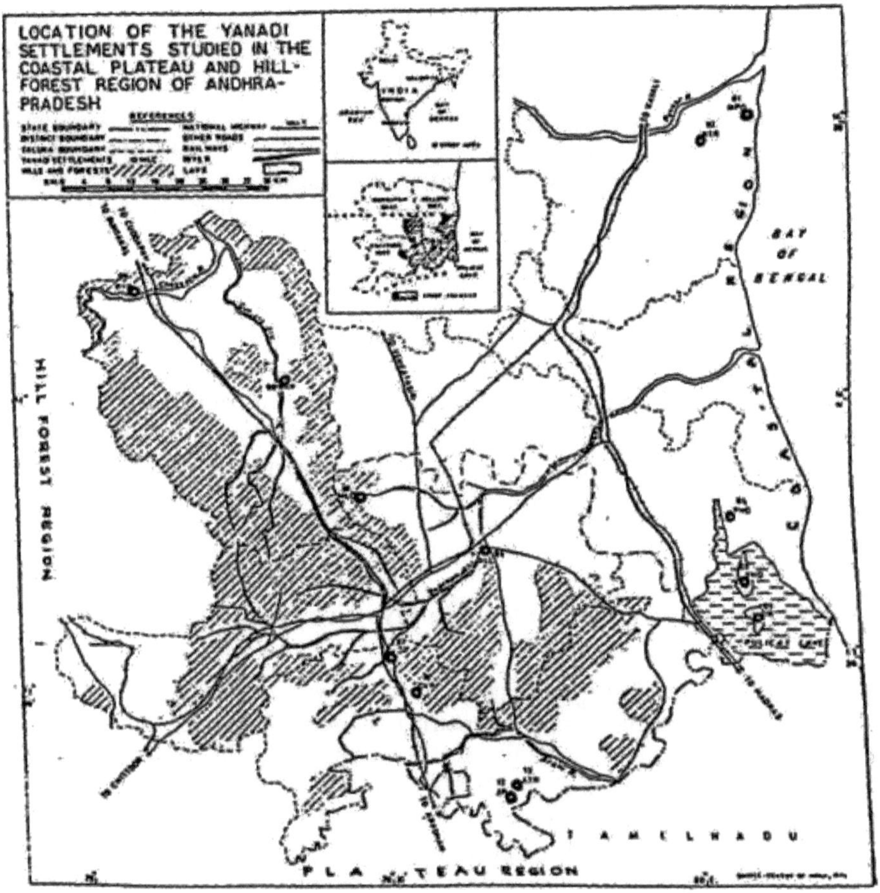

Fig. 1 Location of the Yanadi settlements studied in the coastal and plateau regions in Andhra Pradesh

3.3 *Demography and Population Structure*

3.3.1 Unit of Study and Selection of Settlements

A field study conducted in three districts of Nellore, Chittor and Cuddapah (Kadapa) suggests that the Yanadi live in settlements of varying size, from a few huts to about 100 huts which is associated with different subsistence levels and other cultural variables in different regions of coastal, plateau and hill-forest. In general, as the marriages are contracted among nearby settlements, preferably among relatives, Yandi exhibit several regional breeding populations. The breeding isolation is promoted by proximity of relatives located in the same or nearby settlements, geographical barriers, differences in subsistence pattern and spatial distribution. This has been reflected from a detailed analyses of demography, population structure, endogamy and exogamy, surname analysis, inbreeding levels from a field survey of 13 settlements located in insular, coastal, plateau and hill forest regions. A more details of the methodology and selection of the settlements, identification of regional populations have been described elsewhere (Vasulu 1989). The study identified 5 breeding populations: CY—Challa Yanadi, IY—Insular Yanadi, P1, P2—Plateau Yanadi and HF—Hill forest Yanadi. Detailed demographic, genealogical survey shows that these five breeding populations differ with respect population structure variables. For example, CY and IY form two distinct breeding populations whereas P1, HF and P2 reflect low levels of mate exchange or marital contracts. Details of a few population structure variables especially breeding isolation, admixture rate (m), surname migration index (SMI), surname identity (SII), Crow's Index of opportunity for natural selection (Crow 1958) and effective number of surnames (NeS) of the five regional breeding populations of the Yanadi tribe are given in Table 1. Further details on the breeding populations and population structure of the tribe have been described earlier (Vasulu 1989).

3.4 Anthropometric and Dermatoglyphics Data

For the study, data on 19 anthropometric traits were obtained from men aged 20–60 years, and data on 14 anthropometric traits were obtained from women aged 18–60 years from different settlements of the Yanadi tribe. In general, Yanadi settlements are small in size of about 20–100 individuals. In view of the small size of the nature of the settlements, the study was planned to collect metric data from all the adult individuals from 12 settlements of the tribe, however, in practical no metric data could be obtained from the settlement 9 Hill Forest region and three females could be measured from a settlement (no. 6) in plateau region.

For the analysis of the present study a set of 14 anthropometric measurements which are precise and common to both male and females were considered. These 14 traits are: Stature (HV), head length (HL), frontal breadth (FB), bizygomatic breadth (BZB), biogonial breadth (BGB), exter-canthion diameter (ECD), inter-cathion diameter (ICD), nasal length (NL), nasal breadth (NB), mouth length (ML), mouth breadth (MB), ear height (EH) and ear breadth (EB) Thus the data considered for the study include mostly cephalometric traits as these are moderately or highly heritable characters and are known to reflect population differentiation (Howells 1969; Martin 1928; Friedlander 1975; Lees and Relethford 1982; Weiner and Lourie 1981). Since the settlements are small in terms of size and therefore all the available adults were included for anthropometric data and invariably include related individuals.

Finger and palm prints were collected from all the available individuals, including children, of all the 13 settlements, except in one settlement (no. 6 of P1 region) prints of only three female adult could be taken. The prints were analysed for qualitative and quantitative characters, both of fingers and palms, according to the standard pro-

Table 1 Population structure variables and sample size for anthropometrics and dermatoglyphics characters studied among the five regional breeding populations of the Yanadi tribe

Variable		Regional breeding population				
		CY	IY	P1	HF	P2
Anthropometric traits						
Sample size: male		27	54	80	132	148
Female		28	47	38	75	136
Total		55	101	118	207	284
Percentage of total (adult) population		63.9	45.5	65.2	69.9	84.5
Dermatoglyphics prints (finger and palm)						
Sample size: male		40	68	93	156	171
Female		48	63	48	79	166
Total		88	131	141	235	337
Percentage of total population		52.4	33.5	45.9	46.1	57.7
Population structure						
Admixture rate (Mean + SE):	'm'	0	0	1.2 + 0.8	2.4 + 0.8	0.3 + 0.3
Mean marital distance:	MMD	10.86	12.02	15.81	24.79	18.20
Within regional surname identity:	Sii	0.089	0.142	0.070	0.051	0.060
Effective number of surnames:	Ne(s)	11.13	7.01	14.22	19.34	16.53
Surname migration index:	SMI	17.0	2.5	38.3	11.8	24.8
Inbreeding coefficient:	'f_p'	0.026	0.043	0.029	0.038	0.041
Crow's index—intensity of selection: 'I'		0.904	0.753	0.258	0.567	0.443

cedures. For the study both qualitative and quantitative finger and palm traits have been considered. For the present study seven quantitative characters were considered. These are: FPII (Finger Pattern Intensity Index), PPII (Palmer Pattern Intensity Index), MLI (Main Line Index), "atd" angle (maximum "atd" angle in palms), TFRC (Total Finger Ridge Count), AFRC (Absolute Finger Ridge Count), a-b RC (a-b Ridge count in palms). These traits are moderate to highly heritable characters (e.g., Penrose 1954b, Cummins and Midlo 1943, 1961; Holt 1968; Mukherjee 1966, 1967 etc.) and are least correlated.

3.5 *Analysis*

3.5.1 Anthropometric and Dermatoglyphic Characters

The anthropometric and dermatoglyphic data obtained from men and women were preliminarily screened for outliers, those individuals whose measurements were beyond four standard deviations from the mean were removed for further analysis. There were four outliers which have been discarded. To investigate dermatoglyphic and anthropometric variability, all the characters were standardized with respect to their over all mean and standard deviation, separately for males and females data sets, so that each character will have mean zero and variance one. This allows to investigate variability for both within and between characters and also sex differences among the five regional breeding populations. Although sex differences is not so marked for most of the dermatoglyphic characters, this takes care of the scale differences between sexes, especially in some of the ridge-count data and also effect due to small sample size in some cases. This allows comparing males and females together for within and between populations. Simple descriptive statistics of mean and SD etc. were calculated for the characters for all the settlements or regions. The analysis for within and between population variation were considered at two levels of the study units: (1) trait wise and (2) profile wise. The data were analyzed for within and between populations (and between sexes) for each anthropometric and dermatoglyphic trait and profiles of the five regional populations. The details of the sample size for anthropometric and dermatoglyphic characters are given in Table 1.

The distance analysis, viz., Mahalanobis D^2 and the partitioning into size and shape components was computed based on 14 anthropometric measurements which were precise, well defined and common to both sexes. The distance analysis was computed for the five regional populations, separately for males, females and for sex pooled data and for sex pooled data after standardization. Penrose (1954a) C^2H coefficient based on mean squared distance was estimated for overall anthropometric data between the regional populations of the tribe. Further, analysis of variance technique as per Edward and Cavalli Sforza (1965) method, based on mean square distance matrix computed on the standardized data (Spielman 1973a). To investigate the phenotypic relatedness between regional populations dendrograms or clustering trees were constructed following minimum clustering method.

Spearman's rank correlations (r_s) and simple product moment correlation (r) were calculated for the association between two categories: within sex category [i.e. 10 pairs each for male-male and female-female combinations] and between sex category [i.e., 25 pairs of male-female combinations of the 5 Yanadi regional populations] for each dermatoglyphic and anthropometric characters and also for overall morphological variation and as well for dermatoglyphic profiles. ANOVA test was also considered for the significance of mean differences for each character, separately for males and females based on the original data sets. Similar analysis of variance technique based on the standardized data for the mean square distance (MSD) matrix according to Edward and Cavalli-Sforza (1965) was also applied. They extend the

ANOVA technique to distance matrix for estimating the separate contributions of within and between sex categories to total variability (Spielman 1973a). In this study their technique was also extended to estimate the total variability and its partition to within and between sex categories for each dermatoglyphic and anthropometric character, in which case, instead of distances, the MSD were considered. To get overall dermatoglyphic distances and its two components Size and Shape were also computed (Penrose 1954a). Statistical tests of significance test for the mean differences and equality of percentages have been computed by following the method of Rao (1952).

The nature of the settlement and the marriage patterns among the Yanadi restrict the sample size and the individuals are invariably related to one another which, in statistical point of view, violates the assumption of independency. Therefore, the results of significance tests are likely to be less dependable and should be treated cautiously. The results of the analysis were presented at two different levels: variation in each anthropometric and dermatoglyphic characters and anthropometric and dermatoglyphic profiles depicting variation in overall metric characters for each of the five regional populations. These are considered through graphic representation (seven characters) and tabular representation for total anthropometric and dermatoglyphic variability.

4 Results

4.1 Study Hypothesis

In view of differences observed in case of population structure variables, demography, endogamy etc. among the regional populations of the Yanadi tribe, the following hypothesis can be raised:

1. The population structure variables show differential rates of endogamy among the five regional populations of the Yanadi, and it is expected to show greater homogeneity in anthropometric and dermatoglyphic characters in case of P1, HF and P2 when compared to two coastal populations CY and IY.
2. As the marriage among the Yanadi is patrilocal and patriarchal, a practice where females are contracted from other villages and migrate to husbands' villages after marriages and as a result there will be greater inter village variation among males in anthropometric and dermatoglyphic characters.
3. In view of the patrilocal system practiced among the Yanadi, where both males and females relatively share common genetic endowment, as a consequence, it is expected to show a lesser variation in anthropometric and dermatoglyphic characters between sexes within a region than when compared to between regions.

4.2 Anthropometric Characters

4.2.1 Within Population Variation

Table 2 and Fig. 2 depicts the extent and magnitude of variation in anthropometric traits among males and females of the five regional Yanadi populations. In general, the results suggests two aspects: (a) each character shows distinct pattern and (b) they also reflect sex differences across populations and across traits. To be explicit, HB shows maximum variation in males and NB in females. Least variation is noticed in HL in males and NT in females. Overall, females show greater variability (eight traits) than males (sixt traits). At least three characters showed significant similarity between sexes: a positive association is observed in case of three characters HB, EB (r_s = 1.0 < 0.01) and MB (r_s = 9.0 < 0.05), HL shows insignificant negative association (r_s = −0.6 < 0.05) and the rest 10 traits shows no such similarity (e.g., r_s = 0 in case of HV, NB).

4.2.2 Between Population Variation

Figure 3 shows comparison of sex differences observed in anthropometric characters in case of both within and between the five regional populations of the tribe. There are 25 pairwise differences between sexes, of which the five diagonals marked in square

Table 2 Anthropometric variability in males and females among the Yanadi tribe

Character	Variability (S.D.)		Sex difference
	Male	Female	Correlation coefficient (Spearman's 'r_s')
HV	1.76	1.44	0
HL	0.44	1.25	−0.60
HB	3.68	2.61	1.00*
FB	1.37	1.11	0.10
ZB	1.31	0.89	0.10
GB	1.38	2.79	0.50
ECD	1.79	1.62	0.30
ICD	1.79	1.89	0.70
NL	1.67	1.41	0
NB	0.78	3.38	0.90**
MB	1.93	2.41	0.30
MT	1.57	0.65	0.40
EH	1.15	1.64	1.00*
EB	1.33	1.83	

*p = 0.01 **p = 0.05

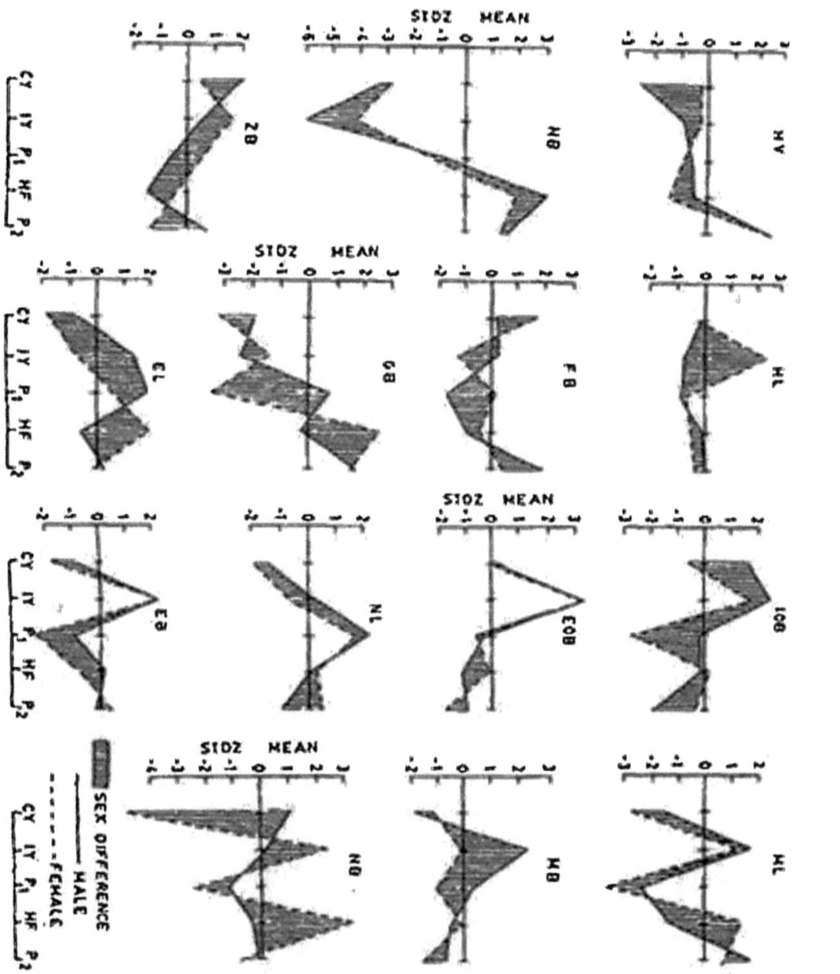

Fig. 2 The extent and magnitude of anthropometric characters in males and females among the five regional populations of the Yanadi

and circle shows the within regions and the rest 20 off-diagonals shows the sex differences compared between regions. Overall, all the 25 figures differ suggesting the pattern of sex differences vary in each case. The figures also show consistence pattern of variation that can be noticed in case of CY, IY with P1, HF and P2 suggesting variation of anthropometric characters in accordance with their spatial distribution of the regional populations and correspond to the differences in population structure variables among the five populations. To be clear, there is a clear tendency of increasing or decreasing trend of mean differences that can be correlated with their spatial distribution, the distantly located population showing greater mean differences—more specifically the two pre-agricultural populations shows higher mean

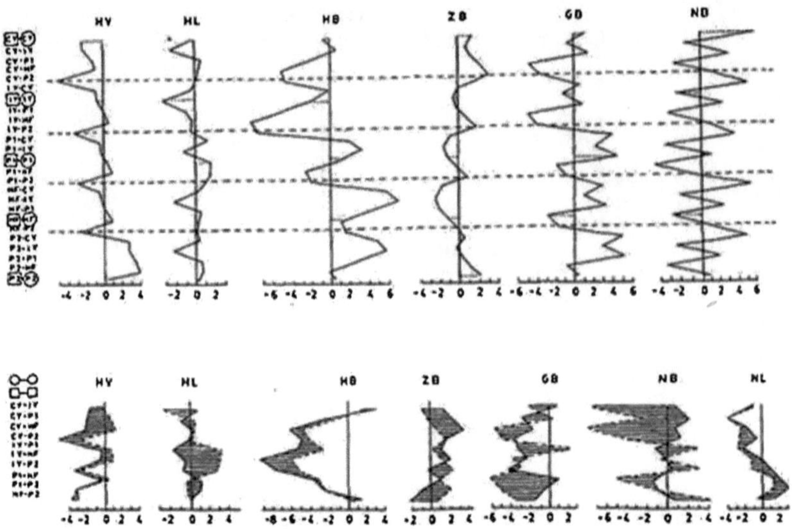

Fig. 3 Between and within sex variation in anthropometric characters among the regional populations of the Yanadi

differences with the three agricultural Yanadi populations. The sharper peaks, in the figure shows greater differences in case of HB, ZB, EH and GB. To a certain extent, similar such trend also can be seen in case of HL. No such differences are noticed in case of three traits: NB, MT, NL. The maximum mean differences are noticed in case of three traits: HB, NB and GB (mean 1.36, 0.37 0.36 and SD 4.06, 3.15 and 3.05).

As it can be inferred from the endogamy rates (one of the population structure variables) the Yanadi regional populations are expected to reflect relatively greater within population metric similarity between males and females. This is inferred from Fig. 3, the five diagonal metric profiles corresponding to within male-females of five regional populations. These figures show a bottleneck shape for the five populations as a result of least average differences observed among males and females within a region. This can be noticed in case of characters: HB, and also in case of ICD, HV, GB as well, though this is not consistent and limited to a few characters. Of the 70 such comparisons involving 14 characters, the least mean differences within regional male and females are noticed in 15 cases (21.4%).

Table 3 and Fig. 3 shows between population comparison of male-female differences in 14 metric characters among five regional populations of the tribe. Overall, the figures shows wide sexual variation in three metric characters: NB, GB and ZB. Maximum variability among males is seen in HB (SD 3.55) and NB (4.36) among females. Females show greater mean differences in 9 (64.3%) traits whereas in males it is observed in 5 characters (35.7%). Does the Table 3 and Fig. 3 indicate an association between the within sex pair combinations? The results of the rank correlation (Table 3) shows significant positive association between male-females in case of 7

Table 3 Overall anthropometric variability between and within sex category among the regional populations of the Yanadi tribe

Trait	Between sex		Within sex				Spearman's Rank correlation
	Male–female		Male–male		Female–female		
	Mean SD		Mean SD		Mean SD		('r_s')
HV	−0.242	2.101	−2.024	1.450	−0.795	1.879	0.600**
HL	0.088	1.279	−0.172	0.861	0.558	1.683	−0.787*
HB	1.359	4.061	−3.808	3.550	−2.887	2.314	0.951**
FB	−0.137	1.619	−0.387	1.909	0.337	1.528	0.127
ZB	0.074	1.448	0.881	1.625	0.948	0.844	−0.551**
GB	0.365	3.152	−1.931	1.640	−2.971	3.222	−0.442
ECD	0.715	2.323	2.076	1.470	0.185	2.289	0.733**
ICD	0.098	2.376	1.351	2.146	1.305	2.303	0.745**
NL	0.305	1.822	0.211	1.967	−1.178	1.610	0.939*
NB	0.367	3.148	0.144	1.086	−1.813	4.356	−0.369
MB	0.285	2.838	−0.643	2.671	−1.457	3.078	0.636**
MT	0.113	1.763	0.132	2.231	−1.190	1.802	0.152
EH	0.626	1.931	0.017	1.677	−1.485	1.776	0.369
EB	0.659	2.153	0.119	1.882	−0.235	2.585	0.793*

*$p = 0.01$; **$p = 0.05$

characters: HB, NL, EB, HV, ECD, ICD and MB. Two characters viz., HL (r_s = −0.79, $p < 0.01$) and ZB (r_s = −0.55, $p < 0.05$) showed negative association and other characters show poor association.

4.3 *Anthropometric/Morphological Profiles*: *Sex Differences*

4.3.1 Within Population Variation

Figure 3 and Table 4 shows male and female anthropometric profiles of the five regional populations of the tribe. Over all, a greater anthropometric variability is observed in case of CY and IY, the two coastal populations than three agricultural populations P1, HF and P2. A majority of the characters show similar profiles between males and females in all the five populations, except in case of NB (e.g., CY and IY), GB (in case of P1 and HF) where females show greater variability. Among the five Yanadi regional populations, significant male-female morphological similarity is observed in case of three populations: IY, P1 and P2 (r_s = −0.56, $p<0.05$).

Table 4 Within sex anthropometric variability between the regional populations of the Yanadi

Male–male				Female–female	Rank correlation
	Mean	SD	Mean	SD	('r_S')
1. Within population					
CY-CY		−1.52		−1.65	0.419
IY-IY		−2.34		−1.94	0.566**
P1-P1		1.24		1.57	0.448**
HF-HF		1.02		1.38	0.316
P2-P2		1.27		1.15	0.478
2. Between population					
CY-IY	−0.971	2.003	−1.827	2.404	0.286
CY-P1	−0.213	2.102	−0.290	1.701	0.513**
CY-HF	−0.410	2.218	−1.983	2.900	0.395
CY-P2	−0.025	2.406	−1.766	2.144	0.645*
IY-P1	0.302	2.766	1.537	2.706	0.588**
IY-HF	0.561	3.184	−0.56	2.624	0.566**
IY-P2	−0.051	3.261	0.061	2.785	0.513
P1-HF	0.069	1.559	−1.695	2.421	0.422
P1-P2	−0.545	2.134	−1.476	2.151	0.395
HF-P2	−0.615	1.567	0.195	1.767	0.084

*$p = 0.01$; **$p = 0.05$

4.3.2 Between Population Variation

A similar anthropometric profiles between males and females were investigated in case of between population (not shown). Of the five populations there are 25 pairs of male-female comparisons. The results show a trend of increasing variability among CY and IY males with females of P1, HF and P2 suggesting that there is morphological differences of the five populations in accordance with the spatial distribution and subsistence differences.

Does the anthropometric profiles indicate within male-female similarity of the five Yanadi populations? The results show within male-female morphological similarity in case of IY and P1 (SD 1.60, 1.44) only. Similar morphological profile similarity was observed in case of off-diagonal figures: especially CY, HF and P2 males show least variability with IY, P2 and HF females. The results of the rank correlation show significant dissimilarity between male-female comparison (off-diagonal) between CY, HF ($r_s = -0.79$), between IY and P1 ($r_s = -0.73$) and P1 and P2 ($r_s = -0.78$).

The morphological profile similarity in case of within sex category (Table 4) also shows similar trend as observed between sex category. For example, among male-male morphological variability IY-HF (SD 3.18) show greater differences. In case of female-female comparison CY-HF show greater variability (SD 2.9), followed by IY-P2 (SD 2.78), IY-P1 (SD 2.71). Least sex differences are noticed among P1-HF and HF-P2 males (SD 1.56), HF-P2 and CY-P1 females (SD 1.72). Is there a morphological similarity in case of within sex combinations? The results of the rank correlations shows significant similarity in anthropometric profiles that can be seen in particular among CY-P2 ($r_s = 0.64$, $p < 0.01$), IY-HF and IY-P2 ($r_s = 0.51, 0.59, 0.56$, $p < 0.05$). However, the three agricultural populations: P1, HF, P2 do not reflect any association with same sex-pair comparisons of HF and P2 populations.

4.4 *Total Morphological Variability*

4.4.1 Variability in Anthropometric Characters

Table 5 shows the male-females differences in 14 anthropometric characters compared in case of within and between regions of the five Yanadi regional populations. The table shows Mean Square Distance and total contribution for each anthropometric trait compared within and between regions Over all, higher values (under total contribution) indicate greater variability of the anthropometric characters among the Yanadi populations. And in a way this corresponds to the wide deviations observed in Figs. 3 and 4. In view of the endogamy rates and patrilineal marriage practiced by the tribe, within the region, male-male category is expected to show least variability than female-female category. Within region, males show least differences in case of seven characters: HL, GB, ICD, NB, MB, EH and EB; NB shows the highest value (MSD 22.6) and higher values are observed in case of 6 traits: HV, HB, FB, ZB, ECD and MT, whereas NL shows near equal values among male-male and female-female categories.

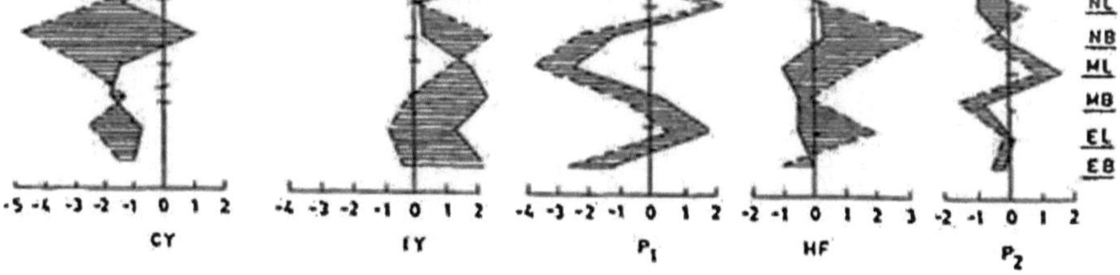

Fig. 4 Anthropometric profiles for males (---) and females (....) of the five regional populations of the Yanadi

Table 5 Total variability in Anthropometric characters due to within sex and between sex contributions in the Yanadi regional populations

Within sex (WS)					Between sex (BS)						
Trait	Male—male		Female—female		Same region		Between region		Total variability		
	MSD	TC	MSD	TC	MSD	TC	MSD	TC	WS	BS	Total
HV	6.19	61.91	4.16	41.65	1.26	6.34	4.98	99.67	103.56	106.04	209.6
HL	0.71	7.71	3.14	31.42	2.52	12.64	1.34	26.70	39.13	145.35	184.5
HB	27.13	271.34	13.69	136.90	1.14	5.69	19.60	392.00	408.24	397.72	805.9
FB	3.19	37.96	2.45	24.50	2.04	10.17	2.64	52.76	62.46	62.93	125.4
ZB	3.42	34.18	1.61	16.09	1.78	9.89	2.07	41.48	50.27	50.37	100.6
GB	6.43	64.32	19.21	192.11	6.02	30.12	9.67	193.57	256.43	236.69	480.1
ECD	6.45	64.52	5.27	52.75	3.22	16.07	5.69	113.97	117.27	130.04	247.3
ICD	6.43	64.28	7.01	70.08	0.42	2.09	6.67	133.97	134.63	135.52	269.9
NL	3.99	39.96	3.98	39.79	0.64	3.18	3.83	76.54	79.75	79.72	159.5
NB	1.19	11.99	22.26	222.61	9.76	48.78	9.46	189.19	234.63	237.97	472.6
MB	7.55	75.48	11.59	115.96	1.99	9.99	9.17	183.49	191.44	193.48	384.9
MT	4.99	49.94	2.43	24.26	2.77	13.86	3.04	60.75	74.20	74.61	148.8
EH	2.81	28.13	5.36	53.62	3.21	16.03	3.69	73.89	81.75	89.92	171.7
EB	3.56	35.55	6.74	67.41	1.53	7.67	5.20	104.06	102.96	117.73	230.7

MSD Mean Square Distance

In accordance with the population structure, within a region male-female differences in anthropometric traits are expected to be relatively lesser when compared to between regions. The results show within region between sex mean differences are least in case of 12 out of 14 traits, the two exceptions are: HL and NB. In case of between regions the male-female difference in some traits, e.g., HB, FOB and HV, EB, MB are about 17 and 14 times higher than within region. In case of within region four traits—NB, GB, EH, and ECD—show higher variability and ICD and NL show the least variation. Whereas in case of between regions, seven characters, viz., HB, GB, MB, ECD, ICD and EB show greater variability and the least variability is seen in case of two traits, HL and ZB. When considered total variability between sex category show higher contribution than within sex category in at least seven characters, viz., HL, FB, ECD, NL, NB, EH and EB, whereas within sex category show higher contribution to total variability in three characters: HB, GB and MB. The traits HV, ZB. ICD, MB and MT show equal contribution from both within and between sex categories.

4.5 *Overall Anthropometric Variability*

A summary of anthropometric variability (based on 14 metric traits) has been considered for within sex and between sex differences across the five regional populations of the Yanadi tribe (Table 6). Since the marriage practice among Yanadi is patrilocal and patrilineal, where brothers in general stay together in a locality and the brides migrate from nearby locations, and as a consequence of which males are expected to show least morphological variability than their wives. The results obtained from the Table 6 does indicate such a trend, where within sex category, males contribution to total morphological variability is nearly five times lesser than female-female contribution. This in conformity with the patrilineal and patrilocal practice of marriage system among the Yanadi tribe. Similar trend is also expected in case of between sex comparison of within and between regions as well. The results show that within regional sex variability is about nine times less than between regions. This conforms to the expectation that with in a region males and females are more similar than between regions.

4.5.1 Association Between Sexes

Figure 5 depicts between sex and within sex anthropometric variability shown as correllelographs across five regional populations of the tribe. The shaded portion of the figures shows sex differences between males and females: the five diagonal figures correspond to the within region and off-diagonal figures is for between regions. The diagonal figures corresponding to the within region male-female reflect higher correlations suggesting morphological similarity. These are highest in case of IY, P2 and P1 and the trend is not reflected in case of CY females and IY males and between

Table 6 Total anthropometric variability due to within sex and between sex populations in the Yanadi

Source	Mean square distance	No. of items	Total contribution
Like sex			
Male–male	2.320	10	23.204
Female–female	7.522	10	75.222
Total	98.426		
Between sex			
Same region	2.736	5	13.681
Between region	6.175	20	123.500
Total	137.181		
Total variability	235.607		

HF females with P2 males. Overall, the figures shows a trend of increasing and decreasing correlations corresponding to their spatial distribution: the farthest shows the least correlations. For example, from CY and IY females showing decreasing correlations with P1, HF and P2 males and similarly, a trend of increasing correlations are noticed in case of HF and P2 females with other males of CY to P2. In case of within sex category of male-male and female-females comparison a significant negative association between females of CY and HF populations was observed. Similar trend is observed in case of IY-HF and CY-P2 populations.

4.5.2 Distance Measure: (a) Size and Shape Components

To get an insight into the overall morphological variability in terms of size and shape, D^2 distance analysis was performed separately for males and females and the results are shown in Table 7. The relative anthropometric similarity/affinity among the five Yanadi regional populations have been examined by clustering pattern (Figure not shown). Overall, shape component (over size) shows wide anthropometric differences among the regional populations. The clustering pattern reflects close anthropometric affinity between the three pre-agricultural inland Yanadi populations, whereas two coastal populations of CY and IY form separate clusters.

4.5.3 Distance Measure: (b) Size and Shape and Sex Differences

To investigate sex differences in overall morphology and its size and shape components, D^2 distance was computed for both sex pooled and sex pooled after standardization and the results are in Table 8. The results show least differences for size component than shape. Similar to Table 7, the three populations P1, HF and P2 show least size and shape differences and cluster together, whereas CY and IY form separate cluster. The dendrograms indicate close clustering of males and females in

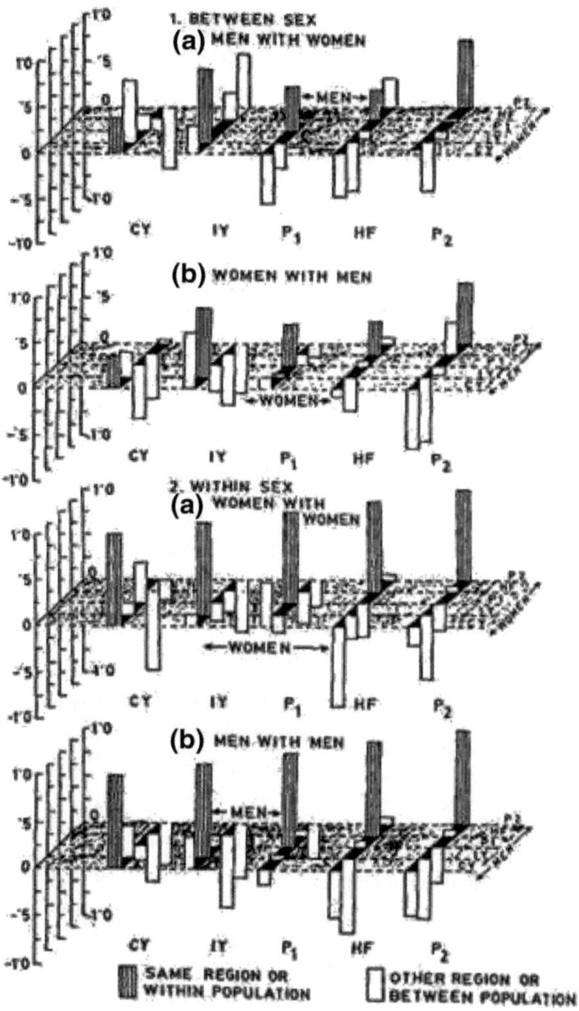

Fig. 5 Correlographs of anthropometric characters among the five regional breeding populations of the Yanadi: 1. Between sex, 2. Within sex (between populations)

case of shape than size clustering. The results are similar in case of standardized sex pooled data also.

4.6 *Dermatoglyphic Characters*

4.6.1 Within Population Variation

The extent of dermatoglyphic variation among males and females of the five Yanadi populations for each of the seven quantitative characters is shown in Fig. 6. The

Table 7 Size and shape components of anthropometric distance (D^2) between the Yanadi breeding populations

Regional populations		Males				
Females		CY	IY	P1	HF	P2
CY	Size	–	0.31	0.30	0.004	0.08
	Shape	–	1.33	1.43	2.56	2.19
	D^2	–	1.54	1.73	2.57	2.27
IY	Size	1.02	–	0.20	1.00	0.11
	Shape	2.11	–	1.93	2.19	2.14
	D^2	3.14	–	2.13	3.19	2.26
P1	Size	0.18	0.10	–	0.04	0.001
	Shape	0.96	3.54	–	0.46	0.85
	D^2	1.14	3.64	–	0.50	0.85
HF	Size	1.18	0.04	0.03	–	0.004
	Shape	2.12	1.94	1.94	–	0.33
	D^2	3.30	1.97	1.97	–	0.34
P2	Size	0.11	0.06	0.10	0.003	–
	Shape	2.15	2.93	1.49	0.63	–
	D^2	2.27	3.00	1.59	0.63	–

Table 8 Size and shape components of anthropometric distance (D^2) between the Yanadi breeding populations (below diagonal—males and females combined), (above diagonal—males and females combined after standardization)

Regional populations		Sex pooled after standardization				
Sex pooled		CY	IY	P1	HF	P2
CY	Size	–	0.49	0.03	0.08	0.34
	Shape	–	0.85	1.28	2.08	1.28
	D^2	–	1.35	1.31	2.16	1.62
IY	Size	0.20	–	0.06	0.05	0.07
	Shape	1.33	–	1.96	2.30	1.84
	D^2	1.54	–	2.03	2.36	1.91
P1	Size	0.16	0.01	–	0.02	0.00
	Shape	1.00	1.92	–	0.50	0.67
	D^2	1.16	1.93	–	0.52	0.67
HF	Size	0.13	0.006	0.001	–	0.002
	Shape	1.81	2.31	0.55	–	0.22
	D^2	1.95	2.32	0.55	–	0.222
P2	Size	0.50	0.21	0.006	0.002	–
	Shape	1.56	1.84	0.70	0.26	–
	D^2	1.86	2.05	0.70	0.26	–

Yanadi show marked differences between the characters in case of sex and region. A greater variability among males is observed for the TFRC (SD 1.49), a-b RC (SD 1.47), AFRC (SD 1.38), "atd"-angle (SD 1.17) and the least is seen in case of PPII (SD 0.66) and atd angle (SD 1.24), FPII (SD 0.93), MLI (SD 0.91) and the lowest in case of AFRC (SD 0.35) among females. Except for "atd"-angle, females consistently show lesser variability than males. The ANOVA test indicate that the differences are significant for TFRC (F 3.13, $p < 0.05$), a-b RC (F 2.81, $p < 0.05$) among males; however females do not show significant differences. Does the magnitude of dermatoglyphic diversity is of the same order between sexes? The rank correlation suggest significant correspondence between sexes for AFRC ($r_s = 1.0$, $p < 0.01$), TFRC ($r_s = 0.90$, $p < 0.05$) and atd angle ($r_s = 0.83$, $p < 0.05$) whereas a-b RC, MLI, FPII ($r_s = 0.0$, -0.10 and -0.3) show very poor correspondence or greater diversity between males and females. The figures suggest that CY and IY together differ from HF and P2 at least in case of atd-angle, AFRC, TFRC and a-b RC.

4.6.2 Between Population Variation

The Fig. 7 shows between sex variability in each of the seven characters among the five regional Yanadi populations. Out of 25 pairs of possible comparison of male-female combinations, the five marked in square and circles representing within regional population variation and the rest 20 pairs represent between regions. The *atd* angle shows the maximum variability (mean 0.13, SD 4.59), PPII shows the least (mean 0.02, SD 0.91) and other show near equal variability (mean ranges from −0.22 to 1.4 and SD from 1.42 to 1.67). The TFRC, AFRC and PPII show minimum fluctuation between CY, IY and P2, whereas MLI and a-b RC show greater fluctuations with CY and IY and least with HF and P2 populations. The Yanadi females of the respective five populations show peak differences with the male populations with respect FPII. Similar peak differences between sexes of a particular region with other regions are seen in case of TFRC, AFRC and a-b RC. Further the curves show a trend of increasing or decreasing differences between populations (and between sexes) corresponding to their spatial distribution and socio-cultural differences, though the differences may not be significant. In particular the trend is observed from CY to P2 populations; especially the female samples of the respective five populations in case of FPII and a-b RC. But in other characters, e.g., PPII, AFRC and TFRC similar gradient between the Yanadi females with the male samples of P1, HF and P2 populations can also be noticed.

Since within a regional endogamous population, both males and females, to great extent, share common genetic and environmental background, it is expected that they show a greater homogeneity in their biological characters than when compared to their counterparts in other regional populations. Accordingly, the dermatoglyphic characters (the curves in Fig. 7) is expected to show least differences (or bottleneck shaped curve) at five places earmarked in squares and circles corresponding to the between sex categories or the five within regional populations. But out of 35 cases of within regions for 7 characters, nearly 50% show the expected least differences, e.g.,

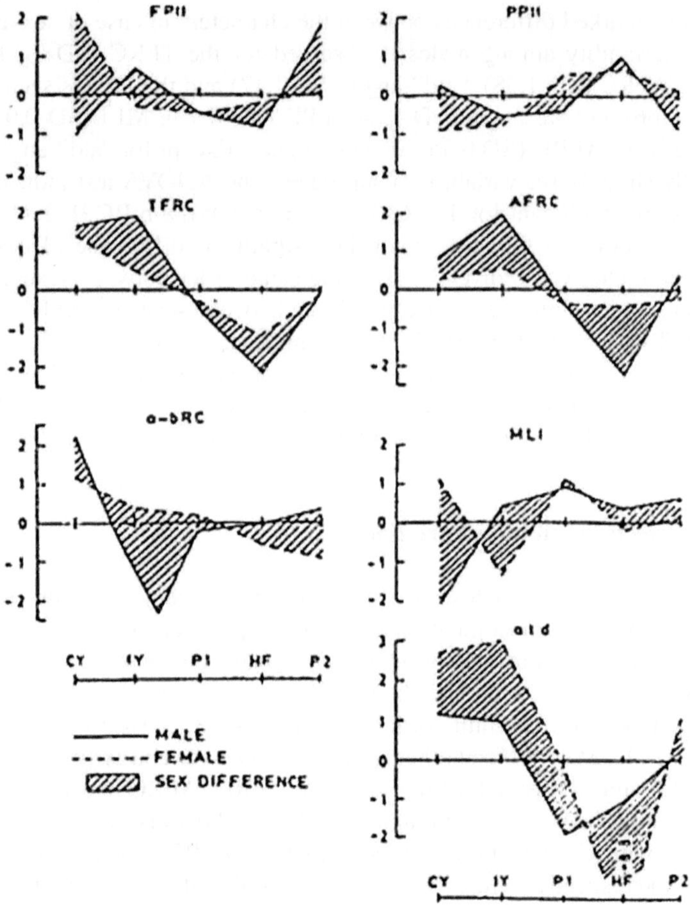

Fig. 6 Dermatoglyphic variation among males and females of the five regional breeding populations of the Yanadi

FPII shows the least differences with respect to P1-P1 and similarly it can be seen in case of PPII trait among HF-HF, in case of atd angle among CY-CY and P2-P2 populations. It is noticed for TFRC among P1-P1, HF-HF, P2-P2 and for a-b RC it is seen among IY-IY and P1-P1. Therefore the hypothesis that males and females within a population show greater genetic homogeneity is not totally inconsistent in case of dermatoglyphics. Further, correlation analysis described later in this paper supports such dermatoglyphic similarity between sexes within populations.

A comparison of within sex variability of dermatoglyphic characters between the five Yanadi populations (Fig. 8) indicate that both male-male and female-female categories show wide differences for the traits FPII (SD 1.46, 1.25), MLI (SD 1.36, 1.41) and a-b RC (SD 2.31, 0.51). The female category shows lesser deviations than males in TRFC, AFRC and a-b RC. The rank correlations for the correspondence

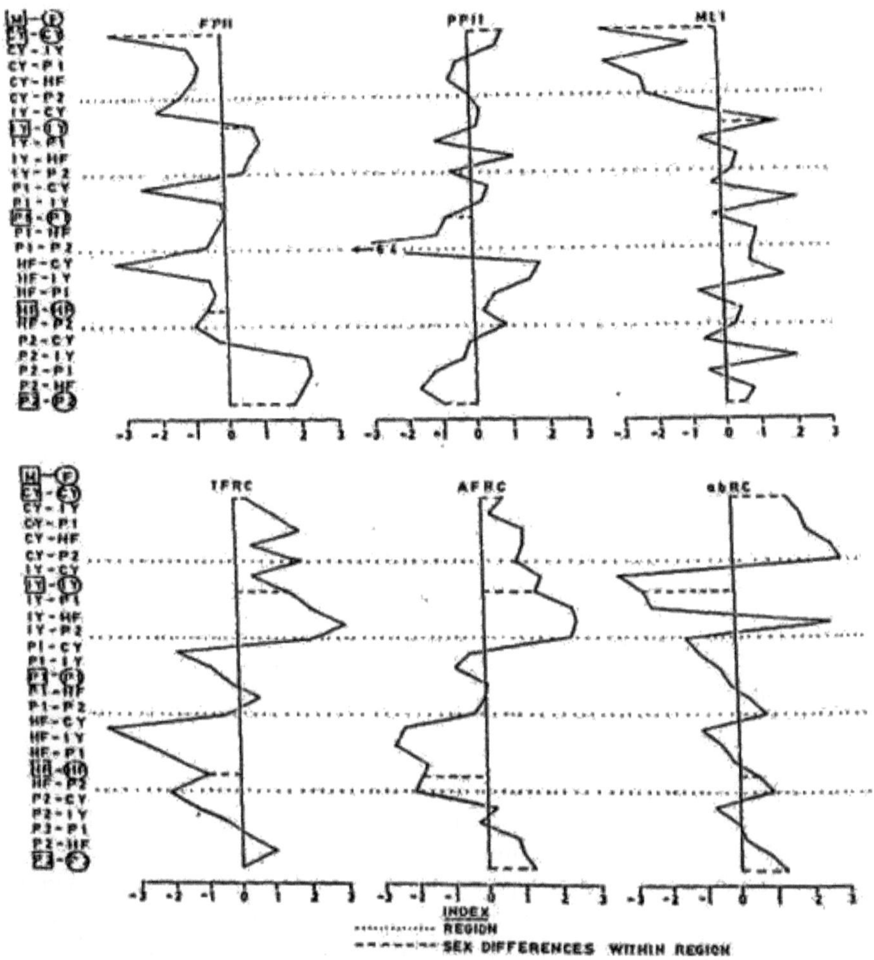

Fig. 7 Between sex variation in quantitative dermatoglyphic characters among the Yanadi—within and between populations

between male-male and female-female categories suggest significant homogeneity in case of *atd* angle ($r_s = 0.64$, $p < 0.05$), TFRC and AFRC ($r_s = 0.74, 0.87$, $p < 0.01$) and poor association (nearly zero) for FPII and a-b RC.

4.7 Total Variability of Dermatoglyphic Characters

The results of the different types of male-female differences in seven dermatoglyphic characters for within and between regions based on analysis of variance are summa-

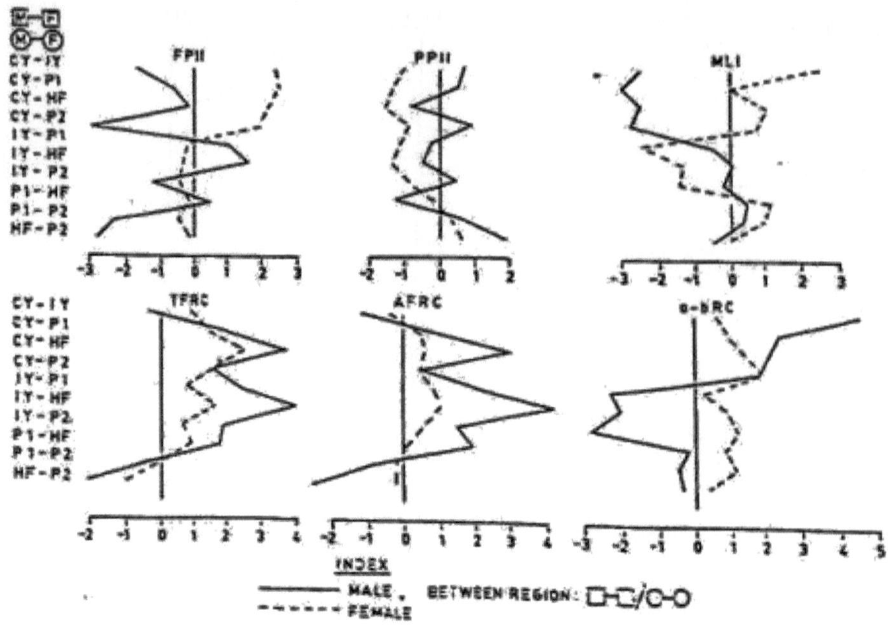

Fig. 8 Within sex variation in quantitative dermatoglyphic characters among the five Yanadi regional populations

rized in Table 9. The table represents summary figures of the matrices of male-female differences in their means for each character among the Yanadi tribe. The higher values observed under the total contribution in the four categories indicate high variability of the characters that correspond with the wide deviations observed in Figs. 7 and 8. In case of within sex category, males show wide variability in all the three ridge counts, whereas it is observed among females for the traits: '*atd*'—angle, FPII and MLI. The least variability is observed in FPII among male and AFRC among female categories, both show about an equal variability for the dermatoglyphic characters FPII and PPII.

In case of between-sex category, males and females of the same region show the least mean differences (when compared to between regions) in 5 characters: PII, '*atd*', TFRC, AFRC and a-b RC. This supports the hypothesis of greater dermatoglyphic homogeneity between sexes within a region. For the total variability, it can be seen from the table that both the within and between sex categories equally contribute in almost all the dermatoglyphic characters except for PPII and TFRC where it is nearly 3 and about twice to between sex category. Similarly, the within regional sex differences are nearly 5 times to between regions in case of PPII and TFRC, and in other characters, the ratio ranges from 0.6 to 1.8 times. Among the Yanadi PPII is the least and '*atd*'-angle is the maximum variable characters.

Table 9 Total variability for the seven dermatoglyphic characters among the five regional populations of the Yanadi tribe: within sex and between sex

Within sex			Between sex			Total variability	
Variable	M–M	F–F	Same region	Between region	Within sex	Between sex	Overall contribution
	TC (10)	TC (10)	TC (5)	TC (10)			
FPII	28.9	21.8	14.4	37.6	50.7	52.0	102.7
PPII	10.8	9.6	2.8	59.4	20.4	62.3	82.6
MLI	30.7	20.3	14.4	38.7	50.9	53.1	104.0
"atd"	34.5	494.2	73.5	455.6	528.6	529.2	1057.9
TFRC	56.2	17.3	3.3	63.0	73.5	66.3	139.8
AFRC	48.0	3.1	6.0	45.6	51.1	51.7	102.8
a-b RC	53.9	11.5	11.0	58.4	65.3	69.5	134.8

TC Total Contribution = *MSD* (Mean Square Distance). number of pairs, *M* Male, *F* Female Figures in the parentheses are the number of pairs

4.8 *Dermatoglyphic Profile*

4.8.1 Within Population Variation

The overall dermatoglyphic differences in all the seven characters together (the dermatoglyphic profiles) among males and females for the five populations are shown in Fig. 9. The profiles show wide fluctuations among males in HF (mean −0.72 SD 1.55), IY (mean 0.43, SD 1.34) and CY (mean 0.38 SD 1.42) and about equal deviation and variability in P1 and P2 (mean 0.38 SD 0.76) populations. Among females both CY and IY (mean 1.12, 0.30 and SD 1.07, 1.28) widely differ from P1, HF and P2 (mean 0.10–0.26 and SD 0.5–0.85). The CY and IY show greater variability when compared to P1, P2 and HF populations, except for atd angle as was earlier observed in Fig. 6. The dermatoglyphic profiles of males and females for the regional populations are similar for most of the characters and are distinct. Though the magnitude of similarity between sexes vary between populations, the degree of association is nearly significant ($p < 0.05$) in IY ($r_s = 0.64$), HF ($r_s = 0.61$) and P1 ($r_s = 0.50$). It is least similar in CY ($r_s = 0.03$) and non similar in P2 ($r_s = -0.25$) populations.

4.8.2 Between Population Variation

The overall dermatoglyphic differences of between sexes in case of within and between regions are shown in Fig. 10. The five rows and five columns represent male-female combinations (between sex categories) for each of the five populations. The diagonal figures earmarked in squares and circles represent within regional and

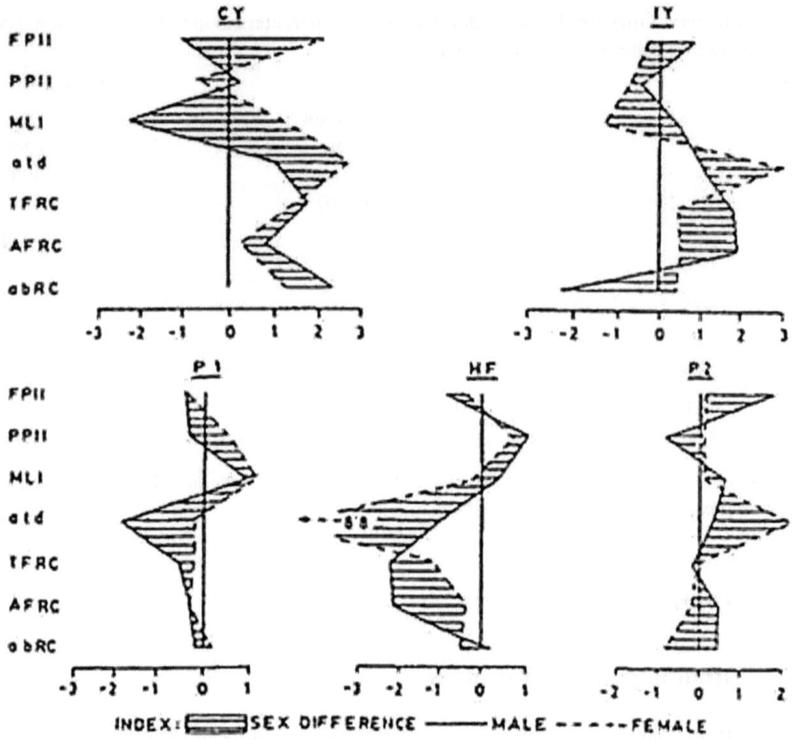

Fig. 9 Dermatoglyphic profiles among the five regional breeding populations of the Yanadi: Males ---, females....

the off-diagonal figures represent between regional sex differences. The results show the least variability for HF-P1 (SD 0.81) and greater dermatoglyphic variability for CY-HF and P2-HF (SD 3.62, 3.00) categories. Is there a correspondence between the male-female differences between populations (i.e., between off-diagonal figures)? The rank correlation suggests significant differences between male-female combinations of HF and P2 ($r_s = -0.75$, $p < 0.05$). In case of within sex category the dermatoglyphic profiles (Fig. 11) males of P1-P2 and females of IY-P2 show least deviations whereas males of IY-HF and females of CY-HF show maximum deviations. Overall, *atd* angle shows the maximum variability in case of females, particularly seen among HF population. When the aberrant *atd* angle is excluded from the analysis, the comparison of between population variability shows least differences among HF, P1 and P2 females. The correspondence between male-male and female-female differences show significant homogeneity in seven dermatoglyphic characters in case of IY-P1 ($r_s = 0.80$, $p < 0.01$) and HF-P2 ($r_s = 0.57$, $p < 0.05$), whereas CY-IY show significant diversity ($r_s = -0.64$, $p < 0.05$).

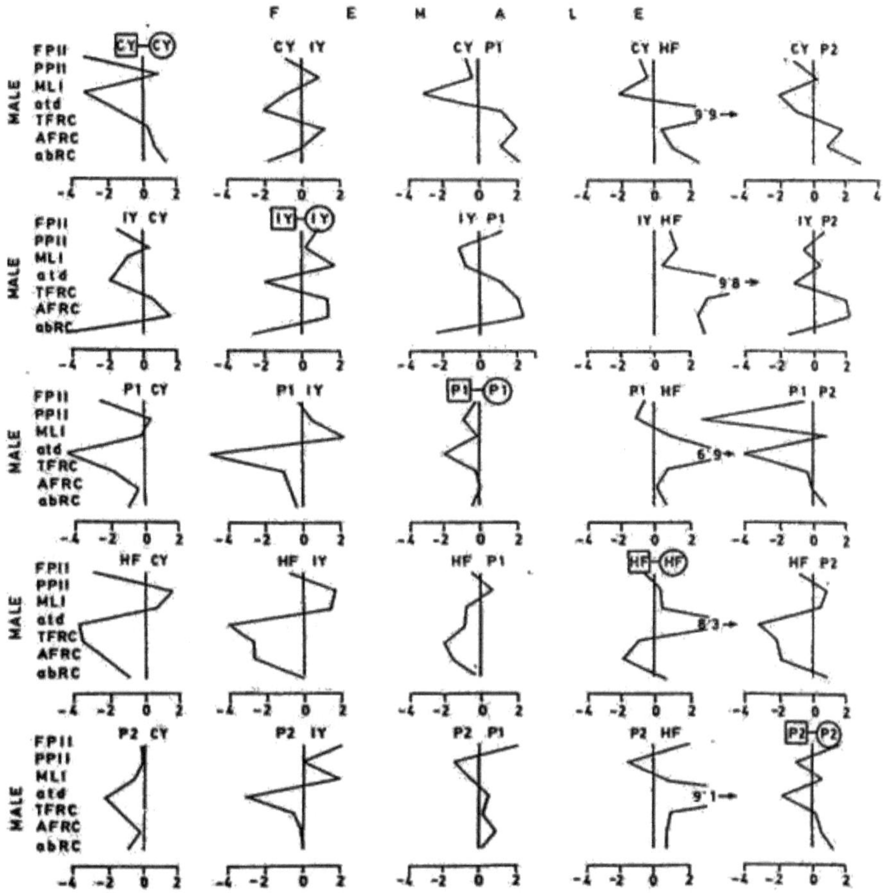

Fig. 10 Dermatoglyphic profiles among the five regional populations of the Yanadi: within populations (diagonal) and between populations (off diagonal)

4.9 *Total Dermatoglyphic Variability*

Total contribution to dermatoglyphic variability due to within and between-sex components based on mean square distance (matrices) are summarised in Table 10. In case of within-sex category, for all the seven dermatoglyphic characters considered the table shows females contribution to total variability is more than the males and from the results obtained in Table 9, it can be inferred that this is contributed mostly by a single character '*atd*'-angle. Similarly in case of between sex category between region contribution to total variability is far greater than the same region category. Therefore, when the analysis is reconsidered based on six characters (without *atd* angle), it is seen that the within sex differences to total mean square distance is about 7 times less than the within male contribution which is in agreement with the results

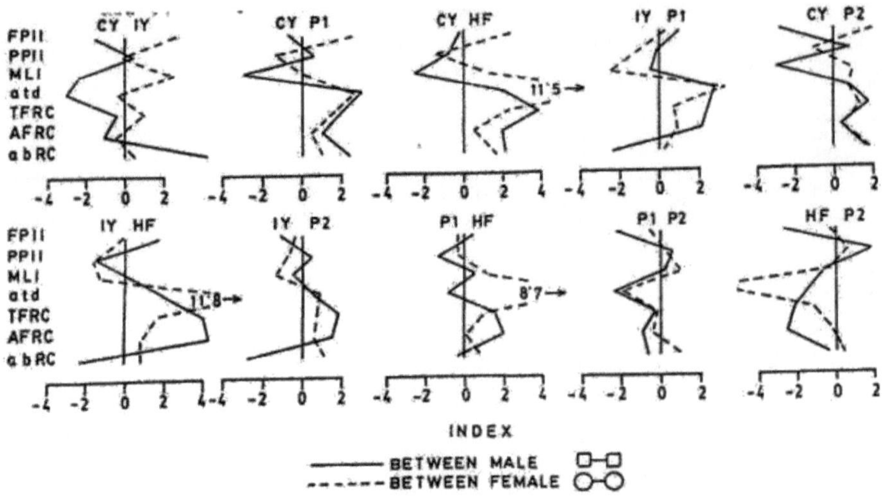

Fig. 11 Dermatoglyphic profiles showing within sex variation among the five regional breeding populations of the Yanadi

Table 10 Total dermatoglyphic variability (six/seven characters) due to within and between sexes among the Yanadi tribe

Source	Six characters[a]			Seven characters		
	MSD	No.	TC	MSD	No. of terms	TC
Like sex						
Male—male	18.1	10	180.6	16.9	10	169.0
Female—female	2.7	10	20.6	137.0	10	1369.7
Total	20.8		207.2	153.9		1538.7
Between sex						
Same region	4.8	5	24.0	22.7	5	113.3
Between regions	6.5	20	129.9	49.2	20	983.1
Total	11.3		153.9	71.9		1096.5
Total variability			361.1			2635.3

[a]"atd" angle is omitted, *MSD* Mean Square Distance, *TC* Total contribution

obtained from individual characters. In case of between sex category, within the same region, the male-female differences are about 5 times lesser than between regional sex differences, suggesting that the males and females of a region show greater similarity in dermatoglyphic characters, when compared to their counterparts in other regions.

4.10 Correlation Analysis

The correspondence between within sex and between sex comparisons in overall dermatoglyphic characters of the tribe in the five studied regional populations are considered in the form of correlographs. Specifically, the correlations between male-male, female-female, and for between sexes among the Yanadi are shown in Fig. 12. The shaded area represents males and females of the respective regions. In case of within-sex category the females of both CY and IY show greater heterogeneity with the other three populations, whereas P1, P2 and HF populations show greater homogeneity among them. In case of between-sex category, the correlations are positive for within the region and the values are significant for IY ($r_s = 0.76$) and P2 ($r_s = 0.68$). The CY males show greater association with IY females. The CY and IY populations show greater association with HF, P1 and P2 populations. In both the cases, decreasing trend of correlations are observed corresponding to one population with the other four regional populations, which is in accordance with the spatial and cultural differences or differences population structural variables; this especially seen between CY, IY with HF, P1 and P2 populations and from Fig. 4 it can be inferred that it is pronounced in case of three ridge count measures.

4.11 Distance Measure

In the dendrograms (Figure not shown) derived from the C_H^2 distances on seven quantitative traits (Table 11) show the same rank order of distances between specific populations for the male sample. The CY and IY populations are located in one extreme and the three agricultural populations tend to cluster together. This pattern differs slightly in case of female data as the P1 population gets clustered with CY and IY populations. The pattern of relationship displayed by these dendrograms and spatial configuration (Fig. 1) are also reflected in that based on C_Z^2 for shape distance (Table 12). The shape distance of quantitative dermatoglyphic traits also show that the P1, P2 and HF cluster together, whereas pre-agricultural populations IY and CY cluster together and are placed apart. It is also noteworthy that even in the female samples (which are less adequate), the CY and IY regional populations show the closest similarity. In case of size distance (C_Q^2) male sample shows close clustering of CY and IY and the three inland populations P1, P2 and HF forming another cluster. The clustering pattern is different among males and females. In case of female sample shows deviation in the sense that CY forms an outlier where as IY population clusters with the three agricultural populations of P1, P2 and HF. The pattern of clustering of C_Q^2 differs from C_Z^2, in males HP, P1 and P2 form a close cluster, whereas CY and IY separates out, CY forming as the outlier. In females both show a single cluster with similar clustering pattern with respect to HF, P1, P2 with CY and IY as deviating from the rest of the populations.

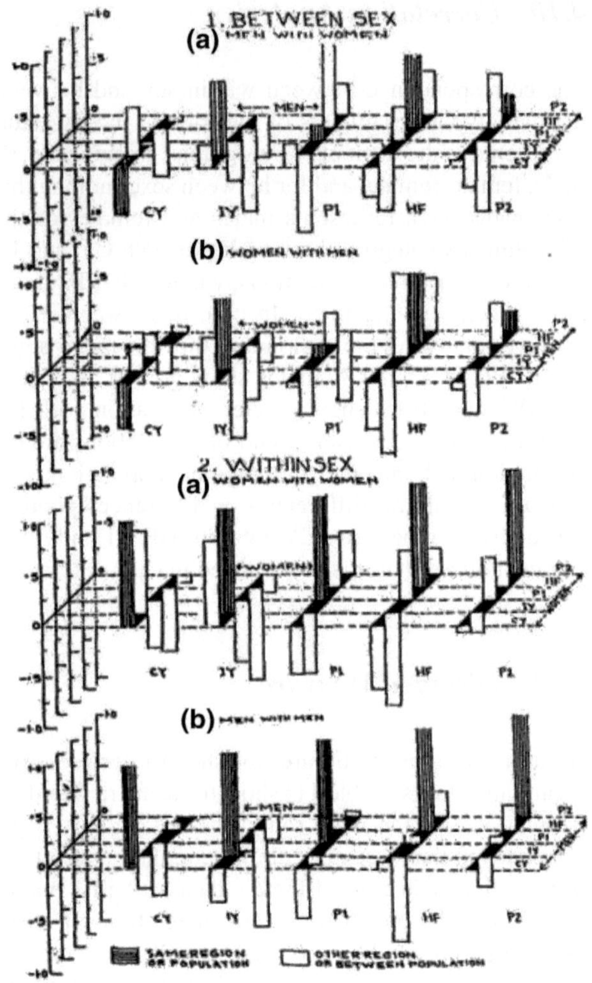

Fig. 12 Correlographs of overall dermatoglyphic variation showing spatio-regional trends in the five regional populations of the Yanadi: 1. Between sex and 2. Within sex

On the basis of these results it can be inferred that there is a pattern in the genetic differentiation, as reflected in case of dermatoglyphic traits, between regional populations, suggesting micro-evolutionary trends among the Yanadi. This pattern or direction of change is consistent with the eco-cultural diversity as well as geographical isolation between the regional breeding populations. The trend of techno-cultural development in the isolated regional populations may have some impact on the microevolutionary changes in the subdivision of the Yanadi tribal population.

Table 11 Dermatoglyphic distance measure (C_H^2) based on seven quantitative traits between the five regional breeding populations of the Yanadi

Regional Populations		Female				
		CY	IY	P1	HF	P2
	CY	–	0.0364	0.1562	0.2280	0.1167
M	IY	0.1895	–	0.0632	0.1235	0.0292
A	P1	0.1789	0.0458	–	0.0526	0.0153
L	HF	0.1882	0.1131	0.0274	–	0.0643
E	P2	0.1808	0.0758	0.0165	0.0144	–

Lower Diagonal—Male, Upper Diagonal—Female

Table 12 Size (C_Q^2) and Shape (C_Z^2) components of distance measure (C_H^2) based on seven dermatoglyphic quantitative traits between the five regional breeding populations of the Yanadi

Regional Populations		Female				
		CY	IY	P1	HF	P2
Size distace—C_Q^2						
	CY	–	0.0121	0.0296	0.0907	0.0388
M	IY	0.0018	–	0.0038	0.0365	0.0075
A	P1	0.0111	0.0039	–	0.0166	0.0006
L	HF	0.0309	0.0178	0.0049	–	0.0108
E	P2	0.0139	0.0057	0.0002	0.0034	–
Shape distance—C_Z^2						
	CY	–	0.0242	0.1229	0.1373	0.0778
M	IY	0.1877	–	0.0599	0.0870	0.0216
A	P1	0.1678	0.0419	–	0.0362	0.0416
L	HF	0.1573	0.0953	0.0224	–	0.0535
E	P2	0.1668	0.0702	0.0164	0.0110	–

Lower Diagonal—Male, Upper Diagonal—Female

4.12 *Pattern Combination Index*

Per cent frequencies of pattern combination index values among the five regional populations are shown in Table 13 . The table show SDM values corresponding to a variety of combinations of patterns that occur on ten fingers of an individual. The Pattern Combination Analysis has four different pattern combinations. This ranges monomorphic, where all the fingers have the same pattern (SDM = 0.0) to dimorphic, trimorphic and tetramorphic pattern combinations (SDM ranging from 0.325 to 1.366). The results of the pattern combination analysis (PCA) is in agreement with the expectations inferred from the population structure of the Yanadi tribe. This confirms, further that when compared to classical method, the PCA method is more powerful in discriminating the subdivided populations. To be specific, IY shows

Table 13 Per cent frequencies of 23 SDM (Shanon Diversity Measure) and Finger Pattern Combination Ratio (FPCR) of mono, di, tri and tetramorphic pattern combinations of four dermatoglyphic traits (Arch, Ulnar Loop, Radial Loop, and Whorl) among the regional populations of the Yanadi

Rank	SDM	FPC		Yanadi (male)					Yanadi (female)				
				CY	IY	P1	HF	P2	CY	IY	P1	HF	P2
		FPC	No. Ratio	39	67	93	151	171	47	63	49	83	165
1	0.000	1	10:0	12.8	7.5	5.4	12.6	12.9	2.1	9.5	6.2	12.0	10.3
2	0.325	2	9:1	25.6	14.9	17.2	13.2	10.5	12.8	6.3	14.3	20.5	13.9
3	0.500	2	8:2	7.7	10.4	21.5	11.3	14.0	23.4	20.6	12.2	20.5	15.2
4	0.611	2	7:3	12.8	13.4	12.9	14.6	18.1	8.5	17.5	18.4	8.4	21.2
5	0.639	3	8:1:1	2.6	4.5	3.2	3.9	3.5	2.1	4.8	2.0	2.4	3.0
6	0.673	2	6:4	30.8	14.9	12.9	22.5	16.9	21.3	19.0	28.6	15.7	12.1
7	0.693	2	5:5	5.1	11.9	12.9	7.9	9.4	17.0	7.9	8.2	8.4	9.1
8	0.802	3	7:2:1	–	1.5	5.4	3.3	4.1	4.3	7.9	4.1	6.0	6.1
9	0.898	3	8:3:1	–	7.5	3.2	3.9	1.2	6.4	3.2	2.0	1.2	3.6
10	0.940	4	7:1:1:1	–	–	–	0.7	–	–	–	–	–	0.6
11	0.943	3	5:4:1	–	5.9	2.2	2.6	3.5	2.1	3.2	–	3.6	0.6
12	0.950	3	6:2:2	2.6	1.5	2.2	1.9	1.8	–	–	2.0	–	–
13	1.030	3	5:3:2	–	–	–	–	1.8	–	–	–	–	2.4
14	1.055	3	4:4:2	–	2.9	–	–	–	–	–	–	1.2	0.6

(continued)

Table 13 (continued)

Rank	SDM	FPC		Yanadi (male)					Yanadi (female)				
				CY	IY	P1	HF	P2	CY	IY	P1	HF	P2
		FPC	No. Ratio	39	67	93	151	171	47	63	49	83	165
15	1.089	3	4:3:3	–	–	–	–	0.6	–	–	–	–	–
16	1.089	4	6:2:1:1	–	1.5	1.1	0.7	1.2					
17	1.168	4	5:3:1:1	–	–	–	0.7	0.6					
18	1.194	4	4:4:1:1	–	–	–	–	–					
19	1.221	4	5:2:2:1	–	–	–	–	–					
20	1.280	4	4:3:3:1	–	1.5	–	–	–					
21	1.314	4	3:3:3:1										
22	1.332	4	4:2:2:2										
23	1.366	4	3:3:2:2										

unique pattern combinations, especially the absence of monomorphic UL, dimorphic UL-RL, and high frequency of trimorphic W-UL-RL frequencies. It also shows least frequencies of PCT-1tr., high frequency of PCT-4tr (correspondingly a low values for MI, MTI and high frequency of T.I., D.T.I., Q.I., and Q.T.I.)., This is attributed to the high rates of endogamy and consanguinity levels observed in IY population. Similar results, especially, least differences in PCT, PCT-tr and PCR. are observed among the three contiguous populations of P1, HF, P2.

5 Discussion

Apart from the Amerindian scenario, human populations in different parts of the world also survive over a variety of geographical and ecological situations with their unique characteristic features that differentiated them from others, e.g., some of these factors are: linguistic, historical, insular, migrant, ecological etc. Several have investigated the biological consequences and the influence of the above factors on the genetic structure of these populations (e.g., Oliver and Howells 1957), among Senoi of Malaysia (Fix 1975, 1978; Fix and Lei-injo 1975; Wang et al. 2007), among Australian aborigines (White and Parson 1973). The Yanadi, a tribal population in transition, live in wide geographical regions and differ in their subsistence economy and population structure variables provide a situation for investigating the microevolutionary trends. The results of the investigation of anthropometric and dermatoglyphic traits among the five regional populations of the tribe, conform to the expectations as proposed based on population structure variables and illustrate microevolutionary trends among the Yanadi tribe.

5.1 Anthropometric Traits

The results of the anthropometric study among the Yanadi shows high variability and sexual dimorphism in some specific characters and least variation in a few characters. Similar results were also observed among other populations as well. What possible explanation can be put forth for differential microevolutionary changes in anthropometric traits, especially that a few characters exhibit greater variability than others. Interestingly, Heirnaux (1972) for similar observation in his study suggested a general explanation: especially that, since human populations live in a wide diverse cultural and ecological environments which promotes rapid changes in some morphological characters than others. Others have given more specific explanation of the influence of some specific factors that include environment and as well genetic influences. For example, Spielman (1973a) observed similar such differential variability in some anthropometric characters among the Yanomama Indians and he attributed to greater genetic influence, despite possible environmental factors. Similar explanation has also been put forth in case of similar such observations among Ireland populations by Lees and Relethford (1982). In case of the present study, among the Yanadi, three anthropometric characters viz., HB, NB, and GB show greater variation among males and females in case of between populations within a region. This could be due to genetic, environmental and both genetic and environmental reasons. In general, some of the anthropometric characters exhibit higher heritability index than others. The twin studies shows higher heritability index for some characters: e.g., HL 0.59, HB 0.61, FB and ZB 0.60, GB 0.71, NL 0.76, EL 0.75 (Clark 1956). Among the Yanadi the greater variation (e.g., HB, ZB, NB) could be due to the genetic influences. In case of Yanadi, high endogamy rates observed is another reason as well. In addition, since the agricultural way of life practiced among HF and P2 is about two or three decades, as such it may not be attributed to the differential variability observed in some characters. The greater variability observed in some specific characters among the Yanadi, could be attributed to population structure variables and high heritability.

Other observation of the study is about sex differences in anthropometric characters. The study showed greater inter-regional variability among males than females. Similar results have been reported among the Yanomama Indians as well (Spielman 1973a). This is explained due to population structure of the tribe, especially patrilocal, patrilineal system practiced, where females after marriage migrate to husband's villages, in contrast, in a region or village, males live together. The consequence of this, at inter village level, it implies greater variability among males than females. The Yanadi tribe is patrilocal, patriarchal and patrilineal. As a consequence, greater homogeneity among males than females in a region is expected and in contrast the males are expected to show greater inter-regional variability. The anthropometric results conforms to the expectation among patrilineal Yanadi. Similar results obtained in case of size and shape components of anthropometric profiles and as well in dermatoglyphic characters supports the influence of population structure variable in case of polygenic traits (Vasulu 1991, 1993). The study emphasizes that the populations governed by similar population structure and living circumstances as observed in the Yanadi tribe in India and Yanomama Indians in Barzil etc., are expected to show similar results of greater within morphological homogeneity and sex differences.

The study showed sex differences in anthropometric characters and morphology that vary between within and between regions. One possible explanation possibly could be sought from the explanation given by Hall and Mac Nair (1972) and Hall (1978). The proposed differential influence of environmental effects between males and females. Especially that males more influenced by environmental factors, on the other hand, females are more adaptive and are expected to show less perturbations to environmental changes as a result of 'canalization', which is expected to result in less morphological variability among females. The possible effect of canalization that might also influence microevolutionary trends among the Yanadi would require further detailed study.

The Yanadi study also shows a trend of increasing or decreasing mean differences in anthropometric characters and as well in anthropometric profiles. Such clinal changes are also observed in case of correlations and correllelographs analysis where the Yanadi tribe show sex differences in case of within and regional populations. For example, the CY and IY females show decreasing correlations in anthropometric characters and in anthropometric profiles with males of CY to P2. The trend of spatial clinal metric variation observed among the Yanadi, possibly suggest' isolation by distance' model of population structure deem suitable in predicting the genetic variability.

The size and shape differences inferred from the selected anthropometric characters indicate wide micro-differentiation at the regional level which is in conformation with the cultural differences and population structure variables of the Yanadi. It is less pronounced among the plateau and hill forest section of the tribe than their counter part population living in island, viz., IY and coastal viz., CY population. Both Challa Yanadi (CY) and insular IY Yanadi are in hunting-gathering and incipient agricultural stage. The varying trend of microevolutionary changes observed in size and shape components of morphology among the five regional populations is in correspondence with differences in their population structure variables and subsistence economy. Of the two components, the populations differ more in case of shape factor than size. The Yanomama Indians also showed same phenomenon. And both male and females are also found to be more similar in shape than in size factor. The results also show that females and males in a regional population are more morphologically similar than males and females between regions.

The dendrograms, though differ in the pattern of clustering between males and females of the five regional populations, overall, there is consistency of clustering pattern with respect CY, IY the two pre-agricultural populations as against P1, HF, P2 which tend form a separate cluster. IY regional population which is an endogamous spatially isolated with high levels of consanguinity and inbreeding levels show significant differences in some of the qualitative and quantitative characters considered. As contrast the contiguous main land populations of HF, P1, P2 show least differences as contrast the two insular and coastal IY and CY populations. This could be expected in view of their population structure variables and geographic isolation factors and as such as a result of microevolutionary changes over generations.

5.2 *Dermatoglyphic Traits*

The dermatoglyphic analyses among the Yanadi show a few marked differences between sex and between regional populations, the explanations of which have to be considered in the background of their spatial distribution and cultural attributes. To be specific, the CY a separate endogamous population of the tribe and IY (insular Yanadi) form separate breeding populations. While both these groups are spatially isolated (Fig. 1), the P1, P2 and HF are contiguous and show low levels of admixture between them. The P1, P2 and HF are spatially nearer to IY than to CY (Vasulu 1989). Given this situation, the wide differences among P1, HF and P2 in a few dermatoglyphic profiles are quite in agreement with the nature of breeding isolation and migration history of Yanadi. Further, the observed trend of increasing or decreasing mean differences form CY to P2 populations, especially seen in three ridge count measures, and the few exceptions to the observed trend, the decreasing correlation in general corroborate with their spatial distribution is also observed in case of anthropometric data as well.

The males consistently show a greater variability of dermatoglyphic differences between populations than female: it is observed especially in TFRC and a-b RC, which are significant in males ($F = 3.13$ and 2.81, $p < 0.05$). Similar observation of significant intervillage variation among males ($F = 4.45$, $p < 0.01$ for TFRC) was also observed among the Yanomama Indians (Rothhammer et al. 1973), where the possible explanation is sought from the patrilocal marriage system. Even among the Yanadi, the marriage is patrilocal and consanguineous i.e., they show greater percentage of wife migration than husband and overall, the wives are related (Vasulu 1989). This in theory should lead to lesser variability among females in case of between populations. The results (Table 2) seem to support the assumption in all the dermatoglyphic characters except *atd* angle.

Further, the remarkable least within regional mean differences is observed in at least five dermatoglyphic characters (FPII, *atd*, AFRC, TFRC and a-b RC) and also in case of overall dermatoglyphic variability. This is in accordance with the degrees of breeding isolation and inbreeding levels, whereby there is a greater homogeneity among males and females in several biological characters, when compared to between regions. Similar results are also obtained in anthropometric characters. There are two exceptions FPII and MLI. However, both show narrow differences to be considered as significant and that warrants further explanation in this regard.

The Yanadi show maximum variability in *atd* angle and least in PPII. The males show significant differences in some characters (TFRC and a-b RC) and females in some other characters (*atd* and FPII). The within-male and within-female population differences show similarity (significant) for the traits AFRC and TFRC and the same trend is also seen in case of dermatoglyphic profiles of three populations: IY, HF and P1. This is also reflected in case of between populations, where males show similar trend of differences with females e.g., atd, TFRC and also observed in profiles among IY-P1 and HF-P2. But in other characters and profiles, there is a trend of dissimilarity in a few characters. In case of within and between populations, the males and females show an opposite trend of mean differences. These findings are not unique to the Yanadi. Similar observations were earlier made, for example, among South American tribes (Rothhammer et al. 1973), Island of Havr (Rudan 1975), among Bulgarians (Karov 1990) etc. Finding a plausible explanation for a few characters (e.g., atd angle) to show marked variability and a few characters (e.g., PPII) least variability needs more detailed study. However, at this juncture, perhaps a general observation made by Hiernaux (1972), especially that changes in human populations differ in time and spare, a few change rapidly and others slow due to situations pertaining to their culture and environmental stress seem relevant to Yanadi populations as well. And equally the same argument seems quite valid to individual dermatoglyphic characters, especially that a few (PPII) are selectively inert (Mayr 1963; Rudan and Schmutzer 1976). The association between dermatoglyphic differences and the geographical distance, which also corresponds with the degree of breeding isolation, supports earlier studies, e.g., in case of Australian Aborigines (White and Parson 1973), Eskimos (Meier 1979; Millicic and Rudan 1991) in illustrating the usefulness of quantitative dermatoglyphic characters for the study of microevolution (Meier 1980). The results of the pattern combination analysis (PCA) supports the results obtained from anthropometric and qualitative and quantitative analysis (Vasulu and Pal 1989; Vasulu 1991, 1998, 1999). The PCA analysis reveals the absence of tetramorphic patterns among females whereas males show a low frequency. The results in general conform to the expectations that in accordance with the patrilineal system males are supposed to show less variation than females in some of the anthropometric and dermatoglyphic characters. However, the PCA analysis show low occurrence of tetramorphic patterns whereas females show its absence, this is not expected. An interesting result needs further investigation. Overall the PCA method confirms that is far efficient and a reasonable method than classical method of treating individual traits separately.

References

Aiyappan, A. (1948). *Report on the socioeconomic conditions of the aboriginal tribes of the province of Madras*. Madras: Government Press.

Amato, R., Miele, G., Monticelli, A., & Cocozza, S. (2011). Signs of selective pressure on genetic variants affecting human height. *PLoS ONE, 6*(11), w27588. https://doi.org/10.1371/journal.pone.0027588.

Basu, A. (1969). The Pahira: A population genetical study. *American Journal of Physical Anthropology, 31,* 399–416.

Blangero, J. (1990a). Population structure analysis using polygenetic traits estimation of migration matrices. *Human Biology, 62,* 27–48.

Cavalli-Sforza, L. L. (1969). Human diversity. In: *Proceedings of the XII International Congress on Genetics. Genetic drift in an Italian population in Readings from Scientific American—Biological Anthropology* (Vol. 31, pp. 405–416). San Francisco: WH Freeman & Company Publishers.

Cavalli-Sfroza, L. L. (1974). Controversial issues in human population genetics. *American Journal of Human Genetics, 26,* 266–271.

Census. (1961). *Village census monograph: Hallimoyar* (Vol. 9. Part IV, 20). Madras: Census of India, 1965.

Census. (1971). Village and Town directory and primary census Abstract, Series II—Parts X: A & B, Nellore, Chittoor and Cuddapah districts.

Chakraborty, R., & Malhotra, K. C. (1981). Dermatoglyphics and genetic distance: A comparative study of variability between populations. *Journal of the Indian Anthropological Society, 16,* 261–269.

Chagnon, N. A. (1972). Tribal social organization and genetic microdifferentiation. In G. A. Harrison & A. J. Boyce (Eds.), *The structure of Human populations*. Oxford: Clarendon Press.

Chagnon, N. A., Neel, J. V., Weitkamp, L., Cershowitz, H., & Ayres, M. (1970). The influence of cultural factors on the demography and pattern of gene flow from the Makirirtare to the Yanomama Indians. *American Journal of Physical Anthropology, 32,* 339–350.

Chai, C. K. (1972). Biological distances between indigenous populations of Taiwan. In J. S. Weiner & J. Huizinga (Eds.), *The assessment of populations affinities in Man*. Oxford: Clarenden Press.

Clark, P. J. (1956). The heritability of certain anthropometric characters as ascertained from measurements of twins. *American Journal of Human Genetics, 8,* 49–54.

Crow, J. F. (1958). Some possibilities for measuring selection intensities in Man. *Human Biology, 30,* 1–13.

Cummins, H., & Midlo, C. (1943, 1961). Finger Print, Palms and Soles. In *An introduction to dermatoglyphics*. South Berlin Mass 01549: Research publishing.
Das, B. M. (1981). *Microevolution*. New Delhi: Concept Publishing company.
Das, B. M. (1985). Microevolutionary trend in northeast Indian tribes: An overview. In Y. R. Ahuja & J. V. Neel (Eds.), *Genetic microdifferentiation in human and other animal populations*.Delhi: Indian Anthropogical Association.
Demarchi, D., & Marcellino, A. (1998). Dermatoglyphic relationships among South Amerindian populations. *Human Biology, 70,* 579–596.
Edward, A. W. F., & Cavalli-Sforza, L. L. (1965). A method for cluster analysis. *Biometrics, 21,* 362–375.
Erikson, R. P., Nerlove, S., Cregor, W. P., & Rommey, A. K. (1970). Comparison of Genetic and anthropological interpretations of population isolates in Aguacatenango Chiapas, Mexico. *American Journal of Physical Anthropology, 32,* 105–120.
Fix, A. G. (1975). Fission-fusion and linear effect: Aspects of the population structure of the Semai Senoi of Malaysia. *American Journal of Physical Anthropology, 43,* 295–302.
Fix, A. G. (1978). The role of kin-structured migration and genetic microdifferentiation. *Annals of Human Genetics, 41,* 329–339.
Fix, A. G., & Lei-Injo, L. E. (1975). Genetic microdifferentiation in the Semai Senoi of Malayasia. *American Journal of Physical Anthropology, 43,* 47–56.
Fleishman, M. L. (1980). An unusual distribution for height among males in a Warao Indian village: A possible effect of lineal effect. *American Journal of Physical Anthropology, 39,* 255–262.
Friedlander, J. S. (1975). *Patterns of human variation: The demography, genetics and phonetics of Bougainville islanders*. Cambridge: Harvard University Press.
Friendlander, J. S., Sgarmella-Zonta, L. A., Kidd, K. K., Lai, L. Y. C., Clark, P., & Walsh, R. J. (1971) Biological divergence in South-Central Bougainville: An analysis of blood polymorphism gene frequencies and anthropometric measurements utilizing tree models, and a comparison of these variables with linguistic, geographic and migrational distances. *American Journal of Human Genetics, 23*, 253–270.
Ganguly, P., & Pal, A. (1972). Secular trend in stature in India. In D. Sen (Ed.), *Indian Anthropology today* (pp. 42–48). Calcutta University.
Glas, B., Sacks, M. S., John, E. F., & Hess, C. (1952). Genetic drift in a religious isolate: An analysis of the causes of variation in blood group and other gene frequencies in a small population. *American Naturalist, 86,* 145–159.
Hall, R. L., & Mac Nair, P. L. (1972). Multivariate analysis of anthropometric data and classification of British Colombian natives. *American Journal of Physical Anthropology, 37,* 401–410.
Hall, R. L. (1978). Sexual dimorphism for size in seven nineteenth century Northwest coast population. *Human Biology, 50,* 159–171.
Harrison, G. A., & Boyce, A. J. (1972). *The structure of Human populations*. Oxford: Clarendon Press.
Hiernaux, J. (1966). Human biological diversity in Central Africa. *Man, 1*(3), 287–306.
Hiernaux, J. (1972). The analysis of multivariate biological distances between human populations: Principles and application to sub-Saharan Africa: In S. Weiner & J. Huizinga (Eds.), *The assessment of population affinities in Man* (pp. 96–114). Clarendon: Oxford.
Holt, S. B. (1968). *Genetics of dermal ridges*. Springfield, Charles C: Thomas.
Howels, W. W. (1966). Population distances: Biological, linguistic, geographical and environmental. *Current Anthropology, 7,* 531–535.
Howels, W. W. (1969). The use of multivariate technique in the study of skeletal population. *American Journal of Physical Anthropology, 31,* 311–314.
Iyer, L. A. K. (1968). *Social history of Kerala* (Vol. I). The pre-Dravidians: Book Centre Publication, Madras.
Jorde, L. B. (1980). The genetic structure of subdivided human populations. In J. H. Mielke & M. H. Crawford *Current developments in anthropological genetics* (Vol. 1, pp. 135–208).

Karve, I., & Malhotra, K. C. (1968). A biological comparison of eight endogamous groups of the same rank. *Current Anthropology, 9,* 109–125.
Karov, G. B. (1990). Asymmetry and intraindividual diversity in the digital dermatoglyphics of Bulgarians. *The American Journal of Human Biology, 2,* 63–73.
Katayama, K., & Toyomasu, T. (1979). A genetic study on the local populations in Mie prefecture IV. The genetic relations among the Kamishima, Toshi, Momotori and Toba populations. *Journal of the Anthropological Society of* Nippon, *87*(4), 377–392.
Kirk, R. L. (1982). Linguistic, ecological and genetic differentiation in New Guinea and the Western Pacific. In M. H. Crawford & J. H. Mielke (Eds.), *Current development in Anthropological genetics 2. Ecology and population structure* (pp. 229–253). New York: Plenum Press.
Kraus, B. S., & White, G. B. (1956). Microdiff evolution in human populations. A study of social endogamy and blood type distribution among western Apache. *American Anthropologist, 58,* 1017–1043.
Kuchemann, F. C., & Lasker, G. W. (1980). Historical changes in the coefficient of relationship by isonomy among the populations of the Otomoor villages. *Human Biology, 51,* 63–77.
Lango, A., et al. (2010). Hundreds of variants clustered in genomic loci and biological pathways affect human height. *Nature 14*, *467*(7317), 832–838. https://doi.org/10.1038/nature09410.
Lasker, G. W. (1946). Migration and physical differentiation: A comparison of immigrant with American born Chinese. *American Journal of Physical Anthropology, 3,* 273–300.
Lasker, G. W. (1960). Migration, isolation and ongoing Human evolution. *Human Biology, 32,* 80–88.
Lees, F. C., & Crawford, M. H. (1976). *Anthropometric variation in Tlaxcaltecans: Prehistory, demography, morphology and genetics* (No. 7, pp. 61–80). Lawrence, Kansas: University of Kansas. Publication in Anthropology.
Lees, F. C., & Relethford, M. H. (1982). Population structure and anthropometric variation in Ireland during 1930s. In M. H. Crawford & J. H. Mielke (Eds.), *Current development in anthropological genetics* (Vol. 2, pp. 385–428). New York: Plenum Publishing Corporation.
Leslie, P. W. (1980). Internal migration and genetic differentiation in St. Barthelemy, French West Indices, In B. Dyke & W. T. Morrill (Eds.), *Genealogical Demography* (pp. 166–177). New York: Academic Press.
Leslie, P. W. (1985). Potential mates analysis and the study of human population structure. *The Yearbook of Physical Anthropology, 28,* 53–78.
Lettre, G., et al. (2008). Identification of ten loci associated with height highlights new biological pathways in human growth *40*, 584–591 (PubMed:18391950).
Livingstone, F. B. (1969). Evolution of quantitative characteristics which are determined by several additive loci. *American Journal of Physical Anthropology, 31,* 355–362.
Livingstone, F. B. (1972). Genetic drift and polygenic inheritance. *American Journal of Physical Anthropology, 37,* 117–126.
Loesch, D., & Lisiewicz, H. (1976). Genetical distance and dermatoglyphic characters. I. Interpopulation distance coefficients. *Human Heredity, 26*(409), 424.
Mahalonobis, P. C., Majumdar, D. N., & Rao, C. R. (1941). Anthropometric survey of United Provinces: A statistical study. *Sankhya, 9,* 90–324.
Majumdar, D. N., & Rao, C. R. (1960). *Race elements in Bengal. A quantitative study*. London: Indian Statistical Institute, Asia Publishing House.
Marquer, P., & Jakobi, L. (1976). Dermatoglyphics and endogamy in Bearn, France. *Man II*, 367–383.
Martin, R. (1928). *Lehrbuch der Anthropologie* (Vol. 1). Jena: Gustav Fischer, Verlag.
Martin, A. O. (1970). The founder effect in a Human isolate: Evolutionary implications. *American Journal of Physical Anthropology, 32,* 351–368.
Malhotra, K. C. (1974). Absence of attached earlobe in four Mendelian isolates. *Journal of Human Evolution, 3,* 429–431.
Malhotra, K. C., Chakraborty, R., & Charavarti, A. (1978). Gene differentiation among the Dhangar caste-cluster of Maharashtra India. *Human Heredity, 28,* 26–36.

Mayr, E. (1963). *Animal species and evolution*. Cambridge: Harvard University Press.

Mielke, J. H., & Crawford, M. H. (1973). *Current developments in Anthropological genetics. Theory and Methods* (Vol. 1). New York: Plenum Press.

Meier, R. J. (1979). Historical and evolutionary explanations for dermatoglyphic diversity among Eskimos of Alaska, Canada and Greenland. In W. Wertelecki & C. C. Platao (Eds.), *Dermatoglyphics fifty years later* (pp. 261–267). New York: Alan R Bliss.

Meier, R. J. (1980). Anthropological dermatoglyphics. *A Review, Year Book Anthropology, 23,* 147–178.

Milicic, J., & Rudan, P. (1991). Population structure of the Adriatic islands—Dermatoglyphic analysis. In B. M. Reddy, S. B. Roy & B. N. Sarkar (Eds.), *Dermatoglyphics today*. Calcutta: IBRAD, ASI and ISI.

Mukherjee, D. P. (1989). Multivariate distance: New perspective and applications in anthropology. In *Changing perspectives of Anthropology in India* (pp. 185–210). New Delhi: Today and tomorrow's Printers and Publishers.

Mukherjee, D. P., & Chakravartti, M. R. (1964). A dermatoglyphic approach to the Mundari ethnic problem. *Zeitschrift für Morphologie und Anthropologie, 55*(1), 32–45.

Mukherjee, D. P. (1978). Changing patterns of marriage and their genetic implications. In: *Proceedings of the 10th International Congress Anthropological and Ethnological Sciences*, New Delhi (Vol. 3, p. 65).

Mukherjee, D. P. (1966). Inheritance of total number of triradii on fingers, palms and soles. *Annals of Human Genetics, 29,* 349–353.

Mukherjee, D. P. (1967). Quantitative genetics of dermal ridges with special reference to an Indian population. Ph.D. Thesis, University of London.

Namouchi, I. (2011). Anthropological significance of dermatoglyphic trait variation: an intra-Tunisian population analysis. *International Journal of Modern Anthropology, 4,* 12–27.

Neel, J. V. (1967). The genetic structure of primitive human populations. *The Japanese Journal of Human Genetics Journal, 12,* 1.

Neel, J. V. (1970). Lessons from a primitive people. *Science, 170*(3760), 815–822.

Neel, J. V. (1978). Rare variants, private polymorphisms and locus heterogzygosity in Amerindian populations. *American Journal of Human Genetics, 30,* 455.

Neel, J. V., & Salzano, F. M. (1964a). Further studies on the Xavante Indians. X. Some hypothesis, generalizations resulting from these studies. *American Journal of Human Genetics, 19,* 554–574.

Neel, J. V., & Salzano, F. M. (1964b). A prospectus for genetic studies of the American Indians. *Cold Springs Harbor Laboratory of Quantitative Biology, 29,* 85–98.

Neel, J. V., & Salzano, F. M. (1967). Further studies on the Xavante Indians. Some hypothesis, generalizations resulting from these studies. *American Journal of Human Genetics, 19,* 554–574.

Neel, J. V., & Ward, R. H. (1970). Village and tribal genetic distances among American Indians and the possible implication for Human Evolution. *Proceedings of the National Academy of Sciences, 65,* 323–330.

Neel, J. V., & Ward, R. H. (1973). The genetic structure of a tribal population. The Yanomama Indians. VI. Analysis by F-statistics (including comparison with the Makritare and Xavante). *Genetics, 72,* 639–666.

Neel, J. V., Rothhammer, F., & Lingoes, J. C. (1974). The genetic structure of a tribal population, the Yanaomama Indians. X: Agreement between representations of village distances based on different sets of characters. *The American Journal of Human Genetics, 26,* 281–303.

Nei, M. N., & Chakrobarti, A. (1977). Drift variances of FST and GST statistics obtained from a finite number of isolated populations. *Theoretical Population Biology, 11,* 307–325.

Nettle, D. (2002a). Women's height, reproductive success and the evolution of sexual dimorphism in modern humans. *The Proceedings of the Biological Sciences/The Royal Society, 269,* 1919–1923.

Nettle, D. (2002b). Height and reproductive success in a cohort of British men. *Human Nature, 13,* 473–491.

Niswander, J. D., Brown, K. S., Iba, B. Y., Leyshon, W. C., & Workman, P. C. (1970). Population studies on south western Indian tribes, I. History, culture and genetics of the Papago. *American Journal of Human Genetics, 22*(1), 7–23.

Oliver, D. L., & Howells, W. W. (1956). Natural selection in modern man. *American Anthropologist, 58,* 591–604.

Oliver, D. L., & Howells, W. W. (1957). Microevolution: Cultural elements in physical variation. *American Anthropologist, 59,* 965–978.

Papiha, S. S. (1983). Genetic structure and microdifferentiaiton among populations of Kinnaur district. Himachal Pradesh. India. In Y. R. Ahuza, J. V. Neel (Eds.), *Genetic mircodifferentiation in human and other animal populations*. Delhi: Indian Anthropological Association (1985).

Penrose, L. S. (1954a). Distance, size and shape. *Ann Eugenics., 13,* 228–237.

Penrose, L. S. (1954b). The distal triradius on the hands of parents and sibs of mongol imbeciles. *Annals of Human Genetics, 19,* 10–27.

Penrose, L. S. (1965). Congenital malformations in man and natural selection. *The Eugenics Review, 57*(3), 126–130.

Rao, C. R. (1952). *Advanced statistical methods in biometric research*. New York: John Wiley and Sens Inc.

Rao, P. R., Sampat, K., Char, N., & Goplan, K. B. (1983). Genetic differentiation in a Brahmin caste-cluster and a Vyasya castes of Andhra Pradesh, south India. In Y. R. Ahuja & J. V. Neel (Eds.), *Genetic differentiation in human and other animal populations.* Delhi: Indian Anthropological Association, 1985.

Raghavaiah, V. (1962). *The Yanadi*. New Delhi: Bharatiya Ademjati Sevak Sangh.

Rakshit, H. K. (1972). Language, culture and race in south India. In S. Sinha (Ed.), *Aspects of Indian culture and society*. *Journal of the Indian Anthropological Society* (pp. 205–232).

Reddy, P. C. (1945). Chencheeta and her divine bride groom. *Eastern Anthropology, 1.*

Reddy A. P., Vijaya Kumar, M., Malhotra, K. C., & Mukherjee, B. N. (1982a). Genetic relationship between the Chenchu, Irula and Yanadi: A reassessment. *Journal of the Indian Anthropological Society, 18,* 31–34.

Reddy, A. P., Mukherjee, B. N., Malhotra, K. C., Das, S. K., & Ramachandraiah, T. (1982b). A serological and biochemical genetic study among the coastal and plateau Yanadis: A tribal population of Andhra Pradesh. *Homo, 33*(4), 174–183.

Relethford, J. H., Lees, F. C., & Crawford, M. H. (1980). Population structure and anthropometric variation in rural western Ireland. Migration and biological differentiation. *Annals of Human Biology, 7,* 411–428.

Roberts, D. F. (1956). A demographic study of a Dinka village. *Human Biology, 28,* 323–349.

Roberts, D. F., & Coope, E. M. (1972). Dermatoglyphic variation in South Midlands. *Heredity, 29,* 293–305.

Rodriguez Florez, C. D., & Colantonio, S. E. (2015). Biological affinities and regional microevolution among pre-Hispanic communities of Colombia's Northren Andes. *Anthropologischer Anzeiger, 72*(2), 141–148.

Rogers, A. (1968). Statistical analysis of the migration component of genetic drift. *American Journal of Physical Anthropology, 77,* 451–457.

Rothhammer, F., Neel, J. V., Rocha, F. D., & Sundling, G. Y. (1973). The genetic structure of a tribal population, the Yanomama Indians. VIII. Dermatoglyphic differences among villages. *American Journal of Human Genetics, 25,* 152–166.

Rothhammer, F., Chakraborty, R., & Liop, E. (1977). A collection of marker gene and dermatoglyphic diversity at various levels of population differentiation. *American Journal of Physical Anthropology, 46,* 51–60.

Rudan, P. (1975). The analysis of quantitative dermatoglyphic traits in the rural population of the island of Havr. *Journal of Human Evolution, 4,* 565–591.

Rudan, P., & Schmutzer, L. (1976). Dermatoglyphic of the inhabitants of the island of Havr. *Yugoslavia, Human Heredity, 26,* 425–434.

Salzano, F. M. (1975). *The role of natural selection in human evolution*. New York: American Elsevier.
Salzano, F. M. (1978). Multivariate studies in tribal societies and human evolution. In R. J. Meier, C. M. Otlen & F. Abdel Hameed (Eds.), *Evolutionary Models and studies in Human Diversity* (pp. 181–199). The Hague: Mouton.
Sanghvi, L. D. (1953). Comparison of genetical and morphological methods for a study of biological differences. *American Journal of Physical Anthropology, 11,* 385–404.
Sanna, S., et al. (2008). Common variants in the GDF5-UQCC region are associated with variation in human height. *Nature Genetics, 40,* 198–203. (PubMed: 18193045).
Smouse, P. E., Vitzthum, V. J., & Neel, J. V. (1981). The impact of random and lineal fission on the genetic divergence of small human groups: A case study among the Yanomama. *Genetics, 98,* 179–197.
Soranzo, S., et al. (2009). Meta-analysis of genome-wide scans for human adult stature in humans identifies novel loci and associations with measures of skeletal frame size. *PLoS Genetics, 5,* e1000445. (PibMed:19343178).
Spielman, R. S. (1973a). Do the natives all look alike? Size and Shape components of Anthropometric differences among Yanomama Indian Villages. *The American Naturalist, 1907*, 694–708.
Spileman, R. S. (1973b) Differences among Yanomama Indian villages: Do the patterns of allele frequencies, anthropometrics and Map locations, correspond? *American Journal of Physical Anthropology, 39*, 461–480.
Spielman, R. S., Da Rocha, F. J., Weitkamp, L. R., Ward, R. H., Neel, J. V., & Chagnon, N. A. (1973). The genetic structure of a tribal population, the Yanomama Indians. VII. Anthropometric differences among Yanomama villages. *American Journal of Physical Anthropology, 37,* 345–356.
Stuart, H. A. (1891). *Census of India* (Vol. I, XIII), 1893. Madras.
Temaj, G., Milicic, J., Behluli, L., Smolej, N., & Tatjana, S. J. (2010). Quantiative dermatoglyphic traits in three population groups of Alabanians from Kosovo. *Anthropologist, 12*(4), 251–258.
Temaj, G., Petranovic, M. Z., Skaric-Juric, T., Behlui, I., Narancic, N. S., Xharra, Sh, et al. (2012). A detection of microevolutionary changes by the analysis of qualitative dermatoglyphic traits: an example of Albanians from Kosovo. *Anthropologischer Anzeiger, 69*(4), 461–472. https://doi.org/10.1127/0003-5548/2012/0209.
Thurston, E. (1909). Castes and Tribes in India. Govt. Press, Madras.
Thurston, E., & Rangachari, K. (1901). Castes and tribes of Southern India (Vol. 7). Government Press, Madras.
Vasulu, T. S. (1989). The genetic structure of a tribal population: 1. Breeding isolation among the Yanadi. *International Journal of Anthropology*, *4*(4), 255–268.
Vasulu, T. S., & Pal, M. (1989). The size and shape components of anthropometric differences among the Yanadi. *Annals of Human Biology, 16,* 449–462.
Vasulu, T. S. (1991). Dermatoglyphic differences among the Yanadi: Quantitative characters. In B. M. Reddy, S. B. Roy & B. N. Sarkar (Eds.), *Dermatoglyphics today* (pp. 322–339). IBRAD, ASI, ISI, Calcutta.
Vasulu, T. S. (1993). Genetic structure of a tribal population: Anthropometric differences between regions and settlements among the Yanadi. *Annals of Human Biology, 21,* 127–144.
Vasulu, T. S. (1998). The method of pattern combination analysis in dermatoglyphics: 1. Diversity and sex differences among the regional populations of the Yanadi. *International Journal of Anthropology, 13*(1), 11–24.
Vasulu, T. S. (1999). Anthropometric variation in the Yanadi: population structure and sex differences. *International Journal of Anthropology, 14*(1), 1–17.
Wang, S., Lewis Jr, C., Jakobsson, M., Ramachandran, S., Ray, N., et al. (2007). Genetic variation and population structure in Native Americans. *PLoS Genetics*, *3*, 2049–2067.
Wang, S. M., & Barrantes, R. (2009). Dermatoglyphic traits of six Chibcha-speaking Amerindians of Costa Rica, and an assessment of genetic affinities among populations. *Revista de Biología (International Journal of Tropical Biología), 57*(suppl. 1), 357–369.

Weedon, M. N., et al. (2008). Genome-wide association analysis identifies 20 loci that influences adult height. *Nature Genetics, 40,* 575–583. (PubMed: 18391952).

Weiner, J. S., & Lourie, J. A. (1981). *Practical human biology*. London: Academic Press.

White, N. G., & Parsan, P. A. (1973). Genetic and socio-cultural differentiation in the aborigines of Arnhem island of Australia. *American Journal of Physical Anthropology, 38,* 5–14.

William-Blangero, & Blangero, S. (1990). Effects of population structure on withingroup variation in the Jirels of Nepal. *Human Biology*, *62*, 131–146.

Workman, P. L., & Niswander, J. D. (1970). Population studies on southern Indian tribes II. Local genetic differentiation in Papago. *American Journal of Physical Anthropology, 22,* 24–49.

Workman, P. L., Leicharelli, P., Agostino, R., Scarabino, R., Sacchi, R., Carapello, E., et al. (1975). Genetic differentiation among Sardinian villages. *American Journal of Physical Anthropology, 43,* 165–176.

Wright, S. (1965). The interpretation of population structure by F-Statistics with special regard to systems of mating. *Evolution, 19,* 395–420.

Yang, J., et al. (2010). Common SNPs explain a large proportion of the heritability for human height. *Nature Genetics, 42,* 565–569. (PubMed:11055368).

A Statistical Study of Platelet Size Decomposition and Related Growth Model

Ratan Dasgupta, Koushik Biswas and Debabrata Dash

Abstract We study the time-dependent distribution of blood platelet size, and its process of evolution with time, when treated with a mobile ion-carrier chemical A23187 that can penetrate into the cells causing size breakdown. The study helps to understand the role of A23187 as a carrier of drugs into cells, and to investigate signaling pathway in platelets that is of relevance in diagnosing diseases like brain stroke, diabetes, etc. It appears that the size distribution of platelets stabilizes after a time period of 3 h and more. Growth pattern of peak and ebb of the time-dependent platelet size distributions is studied to examine the stability of the process in the long run. A proportionate growth model of size breakdown is proposed.

Keywords Blood platelet distribution · A23187 · Proliferation rate · Proportionate growth model

MS Subject Classification: 62P10, 65D10

The original version of this chapter was revised: Authors' names and their affiliations have been included. The correction to this chapter is available at https://doi.org/10.1007/978-981-13-0980-9_12

R. Dasgupta (✉)
Theoretical Statistics and Mathematics Unit, Indian Statistical Institute, 203 B.T. Road, Kolkata 700108, India
e-mail: rdgupta@isical.ac.in; ratandasgupta@gmail.com

K. Biswas
Regional Institute of Ophthalmology, Medical College Hospital, Kolkata 700073, India
e-mail: koushik2907@gmail.com

D. Dash
Department of Biochemistry, Institute of Medical Sciences, Banaras Hindu University, Varanasi 221005, India
e-mail: ddash.biochem@gmail.com

R. Dasgupta (ed.), *Advances in Growth Curve and Structural Equation Modeling*, https://doi.org/10.1007/978-981-13-0980-9_11

1 Introduction

Platelet is a cell which is found in blood, just like red blood cell (RBC) and white blood cell (WBC). When skin is injured or broken, blood cell platelets clump together and form clots to stop the bleeding. The antibiotic chemical A23187 that can cross cell membranes may be used as a molecular tool to make artificial liposomes loaded with anticancer drugs like Topotecan; see, e.g., Tardi et al. (2000). White et al. (1974) report that increase in the cytoplasmic concentration of calcium ions from intracellular sources may be the critical event in triggering platelet contraction and the release reaction. Increased concentrations of cAMP (cyclic adenosine monophosphate) may inhibit the platelet response to A23187 and other aggregating agents by stimulating a calcium extrusion mechanism. Gerrard (1974) observed that high concentrations of ionophore caused the destruction of the platelets on prolonged incubation; while the effects of low concentrations on structure and aggregation reversed completely.

We study the effect of the antibiotic A23187 on human platelets decomposition in relation to both particle sizing and counting by Multisizer 4 particle analyzer machine. Results obtained in the present paper on decomposition process are somewhat different from those obtained earlier.

Size distributions in particle splitting process become bimodal under A23187 administration. Small-sized particles are abundantly generated with diameter near 1 μm due to drug incubation inducing random process in splitting of platelets. Relevance of stochastic process in biology is well studied. Feller (1939) stressed on importance of stochastic process in relation to population growth, and the examples included birth and death process with constant expected birth and death rates. The solution of the case with time-dependent rates was considered by Kendall (1948). Mokaddis (1985) showed that an age-dependent continuous time branching process under some conditions on splitting rates may converge to a compound Poisson process. Stratton and Tucker (1964) proved a similar result for nonhomogenous Markov branching process.

We propose a simple growth model of proportionate growth from elementary consideration of particle splitting under the chemical A23187, in view of the fact that observed processes tend to stabilize over time.

Effect of age on the decomposition is also investigated considering the results obtained from different subjects in three-dimensional analysis on particle concentration in a bimodal density of particle size.

The paper is arranged as follows. In Sect. 2, we describe the methodology in machine-assisted experiment conducted over different time spans and different individuals. Section 3 states the result from the study in terms of density function of platelet size with progress in time, under administration of different chemicals including A23187. Effect of age on size distribution is also studied in three-dimensional representation. In Sect. 4, we propose a simple growth model of proportionate growth, as the process stabilizes. The model puts a cap on the rate of splitting so as to exclude an explosive process. In Sect. 5, we discuss the result obtained from the experiments. Figure legends explain the salient features of the findings.

2 Methodology

A machine called Multisizer 4 from the company Beckman Coulter may count cells in a suspension, and it displays the number of percentage concentration of cells of various sizes and volumes. We consider size distributions of platelets after administering calcium ionophore A23187 of concentration 1 micromole.

In this experiment, blood is collected from individuals and platelets are separated from it, and put in a buffer solution.

The mean platelet size is 2–4 μm. Number of percentage concentrations of platelet of various sizes are then counted, see Fig. 1 for smoothed version of the count, so as to resemble a smooth density function of the concerned variable. The distribution resembles a normal distribution, with mean, median, mode being almost equal, shown in the box inside Fig. 1. The chemical A23187 causes the entry of calcium inside the platelet, and breaks down the platelets into small particles. Stability of size distribution of platelets under splitting in the long run is of interest in this study.

To study the stability of the realized process over time, we incubate the platelet suspension with calcium and ionophore A23187 for 15 min then take the reading of the size distributions in number, percentage of cells versus cell size. In another sample, we take platelets, calcium and DMSO, which is a solvent of A23187, incubate it for 15 min, and take reading of the size distributions in number, percentage of cells versus size of cells and this act as a control in the experiment. The experiment is repeated over different subjects of different ages with duration of the experiment as 15 min. Variation of the process over different individuals is also of interest. The study is repeated over individuals of different age so as to examine association of the breakdown process over age of individuals. In the present experiment, the process is studied at different time points and continued up to an hour for two individuals, and continued up to 3 h for an individual to check for stability of size distribution in the long run.

In the graph, resting platelet (RP) denotes normal platelets.

3 Results

We see that on incubating platelets with DMSO and calcium, there is almost no change in platelet size distribution, but when we incubate normal platelets with the chemical A23187, change in size distribution is observed.

All blood samples are analyzed in the same machine. The blood samples are from different individuals for different sets of experiments, e.g., 06-02-13-M30 RP denotes that the experiment was done on 06-02-13, 'M' stands for male and the number after M denotes the age of the person in years, RP denotes resting platelets (normal platelets), and writing after RP denotes the name of other chemicals added in the experiment. The experimental setup is the same except for different chemical treatments. The machine is standardized by the company which has manufactured it. The smoothed curve helps to reveal the growth pattern out of random jig-jag fluctuations in particle readings.

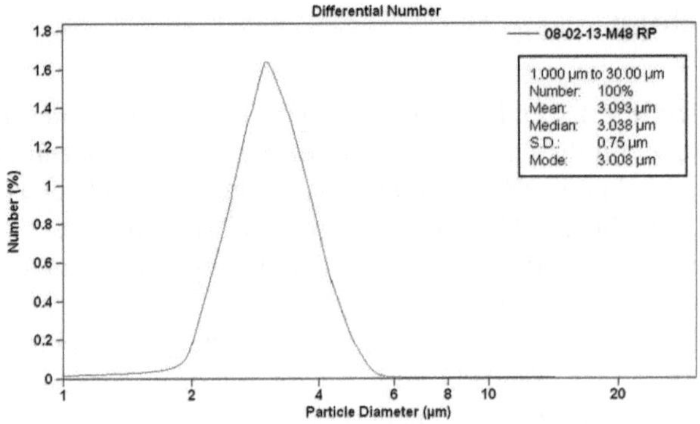

Fig. 1 Density of rest platelet size (M 48, 08-02-13). This figure shows a symmetric unimodal distribution of blood platelet diameter size at rest position for a 48-year-old male, when the platelets are in its natural state, without subjecting these to any chemical interaction. Size range of platelets is [1, 30] microns (μm). Mean, median, and mode of the distribution are all near to 3 μm. We considered the density function of the variable to study size pattern. In continuous distributions, unimodality can be explained through the behavior of the cumulative distribution function, see, e.g., Khinchin and Ya (1938). Size distribution in this figure appears to be normal. Observations are taken on 8 February 2013 on a Multisizer 4 machine. The scale at base is not equispaced

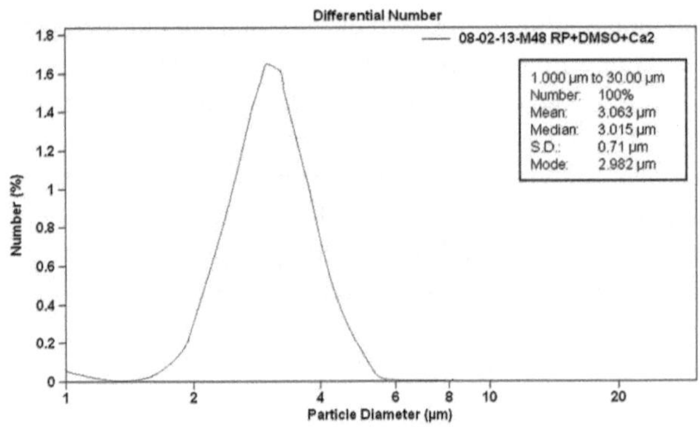

Fig. 2 Density of platelet size after 15 min with DMSO (M 48, 08-02-13). This figure shows platelet size distribution of the 48-year-old male when in a sample we separate platelets from blood, put this with calcium and DMSO, which is a solvent of the chemical A23187, incubate it for 15 min, and then take the reading of the size distributions in number expressed in percentage of cells versus size of cells. DMSO acts as a control in the experiment. Size distribution remains almost unchanged; the value of the parameters mean, median, mode, etc., shown in the box is similar to that of previous figure, indicating the absence of interaction of platelets with calcium and DMSO present in the solution

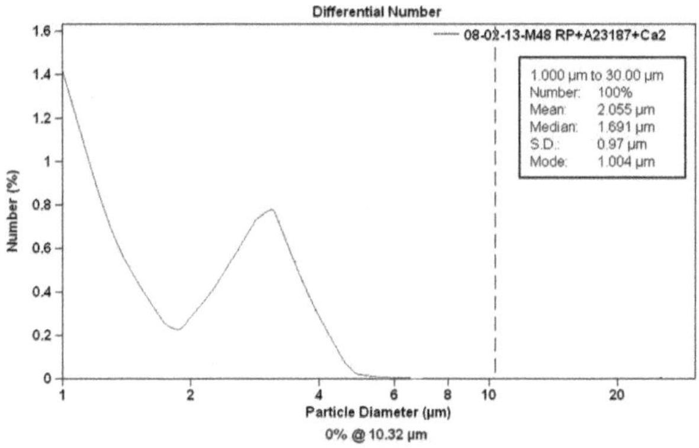

Fig. 3 Density of platelet size after 15 min with A23187 (M 48, 08-02-13). The situation changes when an active chemical agent A23187 is incubated for 15 min with calcium in the platelet solution. Platelets of higher sizes break down to lower values and there is an abundance of small size particles generated concentrating in the solution. The commonly referred bimodal density has a peak at 1.0004 μm. The other mode is at 3.143 μm. There is no particle with diameter above 10.32 μm. The parameters, mean, mode with highest density, median, etc., are shown in the box of this figure

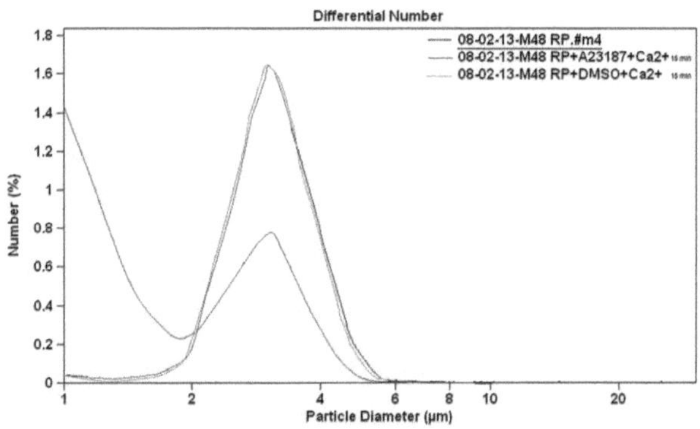

Fig. 4 Density of platelet size with different chemicals (M 48, 08-02-13). In this figure, we show Figs. 1, 2, and 3 in the same frame for easy comparison. Black and green curves corresponding to rest platelet distribution, and distribution under DMSO administration respectively, are overlapping with each other, confirming that no or little change in distribution with DMSO incubation from rest state. However, there is a significant change in size distribution with A23187 administration shown in the brown curve. All observations are recorded in Multisizer 4 machine (#m4)

In Fig. 2, from a close lookup, it is seen that a part of the size distribution is slightly shifted towards left, i.e., towards 1 μ_m under DMSO compared to RP, when unimodal size distribution is observed; mean, median, and mode of the distribution provided inside Fig. 2 do not change much from those given in Fig. 1.

When treated with A23187, the size distribution becomes bimodal with two local maximum, the distribution can be interpreted as a mixture of two distributions; values of mean, median, and modes undergo drastic changes. The mixture proportion is about 0.5. Two modes of the changed distribution (with A23187) are at 1 μ and 3.5 μm approximately, see Fig. 3.

These broad features of change in distribution from RP state are more or less observed in subsequent graphs as well. Tiny sized platelet particles (near 1 μm) are abundantly generated with A23187 administration.

Mode of changed distribution, observed as peak in the graphs, varies over different graphs; mainly for A23187. This also varies for RP when different individuals of different ages are considered.

Care should be taken to interpret the graphs as the scale on base is nonhomogeneous.

The process seems to stabilize after 3 h. The distribution is bimodal for each time point after A23187 administration. There is a little or no change of distribution under DMSO administration. The situation changes drastically after incubating with the chemical A23187. Incubation with DMSO has little change in size distribution in Fig. 2. Observing the pictures in the same frame of Fig. 4 explains the change in distributional pattern. Experiments are carried out with different individuals of different ages in Figs. 5, 6, 7, 8, 9, 10 and 11. These pictures indicate concentration of small

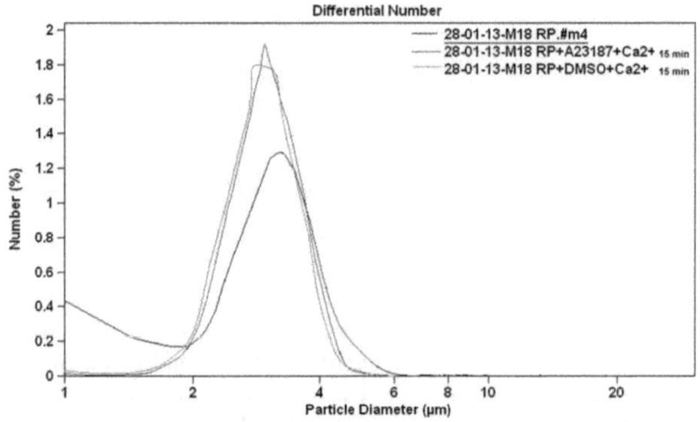

Fig. 5 Density of platelet size with different chemicals (M 18, 28-01-13). This figure shows size distribution of platelets from a young 18-year-old male on 28 January 2013, when chemicals are added in platelet solution, and incubated it for 15 min, and then readings of the size distributions are taken on number in percentage of cells versus size of cells. Black and green curves corresponding to rest platelet distribution and distribution under DMSO administration are overlapping with each other, confirming that no or little change in distribution with DMSO incubation compared to rest state. There is a significant change of size distribution with A23187 administration, shown in the brown curve. This curve is different from the corresponding curve of Fig. 4. One mode of brown curve shown here is near 3.57 μm. Platelet concentration of the 18-year-old male near size 1 μm, under A23187 incubation is less compared to that shown in Fig. 4, corresponding to a male individual aged 48 years

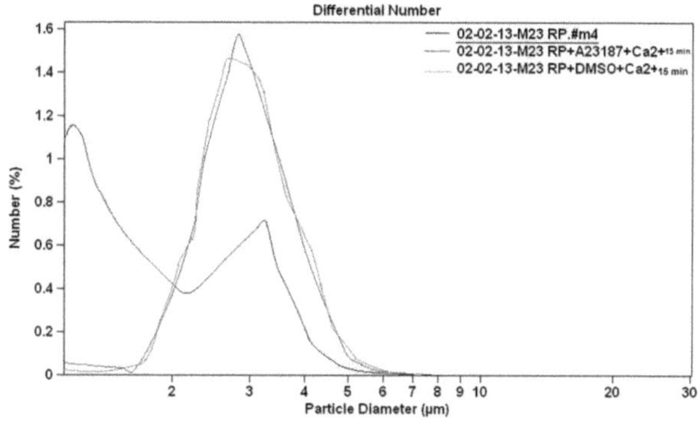

Fig. 6 Density of platelet size with different chemicals (M 23, 02-02-13). This figure shows size distribution of platelets from a slightly higher aged individual than the subject in previous experiment. A 23-year-old male is examined and platelet count is recorded on 2 February 2013, chemicals are added in platelet solution, and incubated it for 15 min, and then we take the reading of the size distributions in number, i.e., percentage of cells versus size of cells. Black and green curves corresponding to rest platelet distribution and distribution under DMSO administration are overlapping with each other, confirming that no or little change in distribution with DMSO incubation has occurred. There is a significant change of size distribution with A23187 administration shown in the brown curve. This curve is different from the corresponding curves of Figs. 4 and 5. One mode of brown curve shown here is near 1 μm, with density at mode being lower than that in Fig. 4

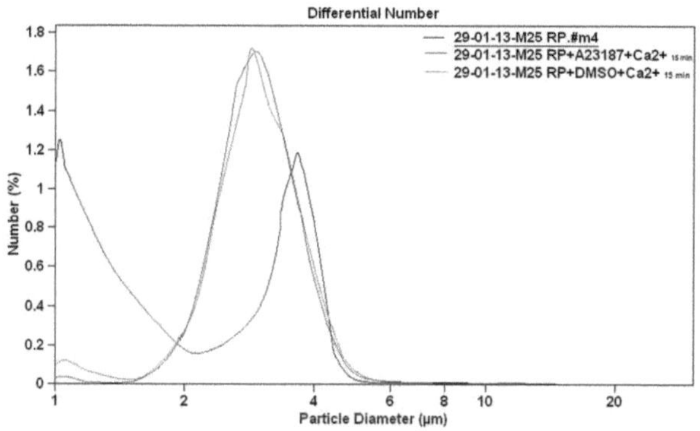

Fig. 7 Density of platelet size with different chemicals (M 25, 29-01-13). This figure shows size distribution of platelets from a 25-year-old male recorded on 29 January 2013, when chemicals are added in platelet solution, and incubated it for 15 min, and then we take the reading of the size distributions in number, i.e., percentage of cells versus size of cells. Black and green curves corresponding to rest platelet distribution and distribution under DMSO administration are overlapping with each other, indicating no or little change in distribution with DMSO incubation. There is a significant change of size distribution with A23187 administration that usually result in a bimodal density, shown here in the brown curve. This curve is different from the corresponding curves of previous figures. Two modes of brown curve are near 1 and 4 μm, with modal density almost equal for two modes

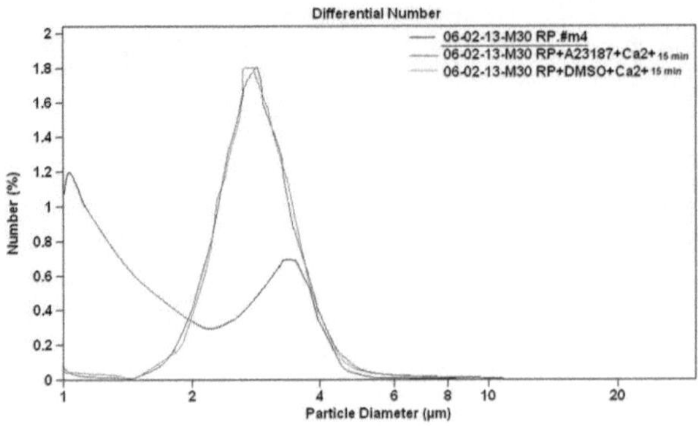

Fig. 8 Density of platelet size with different chemicals (M 30, 06-02-13). This figure shows size distribution of platelets from a 30-year-old male recorded on 6 February 2013, when chemicals are added in platelet solution, and incubated it for 15 min, and then readings of the size distributions in number, i.e., percentage of cells versus size of cells are recorded. Black and green curves corresponding to rest platelet distribution and distribution under DMSO administration are overlapping with each other, confirming that little or no change in distribution has occurred with DMSO incubation. There is a significant change of size distribution with A23187 administration shown in the brown curve. Two modes of brown curve shown here are near 1 and 3.571 μm

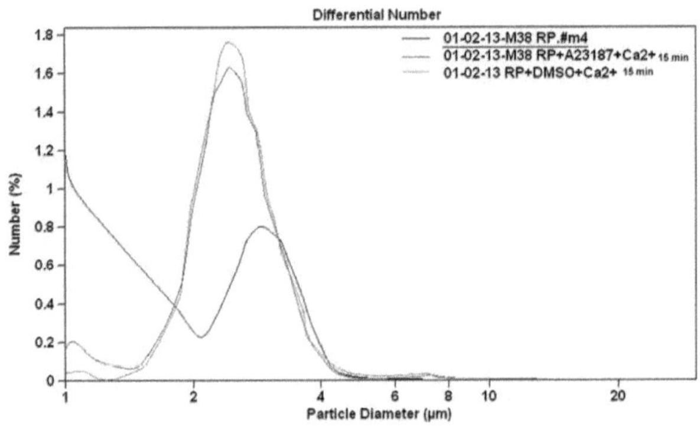

Fig. 9 Density of platelet size with different chemicals (M 38, 01-02-13). This figure shows size distribution of platelets from a 38-year-old male recorded on 1 February 2013, when chemicals are added in platelet solution, and incubated it for 15 min, and then readings of the size distributions in number, viz., percentage of cells versus size of cells are taken. Black and green curves corresponding to rest platelet distribution and distribution under DMSO administration are overlapping with each other, indicating that no or little change in distribution with DMSO incubation. There is a significant change of size distribution with A23187 administration shown in the brown curve. Modes of the bimodal brown curve shown here are near 1 and 3.091 μm

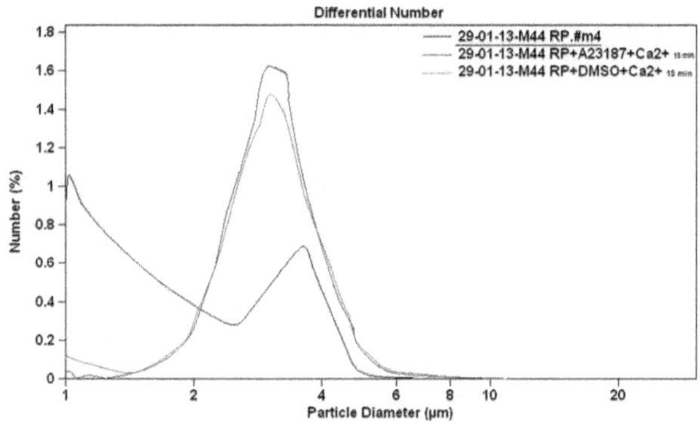

Fig. 10 Density of platelet size with different chemicals (M 44, 29-01-13). We are studying change in pattern of platelet densities for increasing age of subjects with respect to chemicals used in incubation. This figure shows size distribution of platelets from a 44-year-old male on 29 January 2013, when chemicals are added in platelet solution, and incubated it for 15 min, and then readings of the size distributions in number, percentage of cells versus size of cells are taken. Black and green curves corresponding to rest platelet distribution and distribution under DMSO administration are overlapping with each other, indicating no or little change has occurred in distribution with DMSO incubation. There is a significant change of size distribution with A23187 administration shown in the brown curve. Two modes of brown curve here are near 1 and 3.571 μm

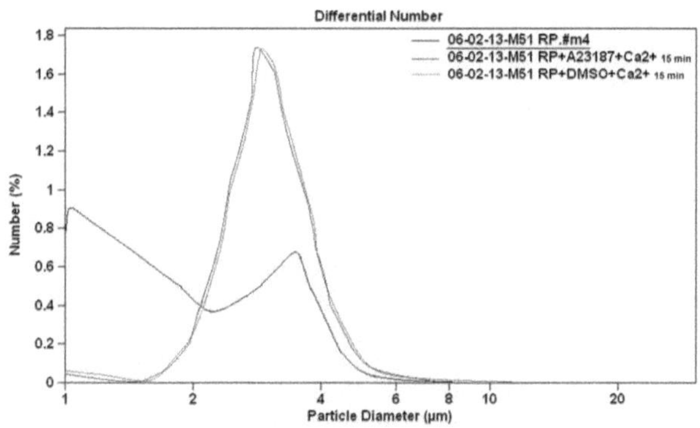

Fig. 11 Density of platelet size with different chemicals (M 51, 06-02-13). The subject is 51-year-old male, having highest age in the study group. This figure shows size distribution of platelets from this individual on 6 February 2013, when chemicals are added in platelet solution, and incubated it for 15 min, and then readings of the size distributions in number and percentage of cells versus size of cells are taken. Black and green curves corresponding to rest platelet distribution and distribution under DMSO administration are overlapping with each other, confirming no or little change in distribution has occurred with DMSO incubation. There is a significant change of size distribution with A23187 administration shown in the brown curve. Two modes of brown curve here are near 1 and 3.429 μm

size cells in the platelet solution after 15 min of interaction. As already mentioned, DMSO$^+$Ca2$^+$…, A23187$^+$Ca2$^+$…, etc. refer to compositions of chemicals.

Longitudinal studies on size distribution are conducted on different individuals over time. These indicate gradual shift of the second mode towards right and increase in density near the first mode at 1 μm as seen in Figs. 12, 13 and 14. A vertical line at 1.5 in Fig. 13 indicates 0.00338% of platelets are below platelet size 1.435 for resting platelets.

Lowess regression, see, e.g., Cleveland (1981) and spline smoothing, see, e.g., Reinsch (1967) are two nonparametric techniques for fitting smooth curves to given data. Spline regression on the set of values of second mode, the peak of density; and the ebb, the lowest value of density near the left side of figures indicate a possible stability of the chemical process of size breakdown. The differences of platelet diameters at the peak and ebb values indicate the same as the distance approaches to a constant over time. Three-dimensional figures with the variables "time", "density height" as number (%), and "particle diameter" in longitudinal studies identify the region of stability of the process.

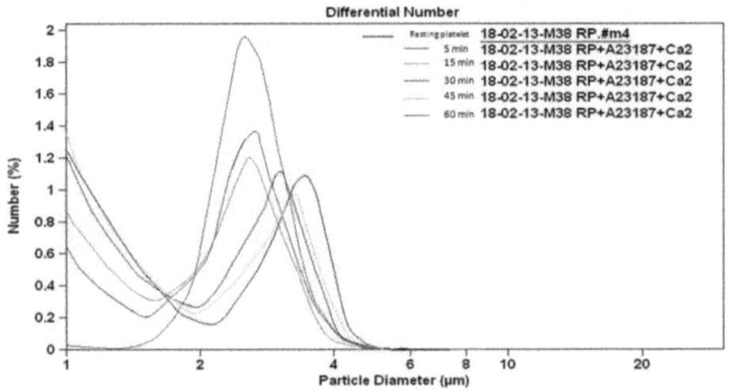

Fig. 12 Density of platelet size with A23187 (M 38, 18-02-13). We now consider longitudinal study on platelets under A23187 treatment over a time period of 1 h recorded on 18 February 2013. The subject is a 38-year-old male. This figure shows size distribution of platelets observed, when the chemical A23187 is added in platelet solution, and incubated for 5, 15, 30, 45, and 60 min, and the readings of the size distributions in number, viz., percentage of cells versus size of cells are taken at the specified time points. Deep blue curve corresponds to rest platelet distribution. The curves are drawn over increasing time from start at rest platelet (RP) position. A part of the curves sequentially shift to the left showing concentration of smaller sized platelets towards 1 μm diameter, indicating massive breakdown of the large platelets caused by A23187. Note the gradual shift to right, for second mode peak, and movement of ebb in left-hand side of the curves. There is a significant change of size distribution with A23187 administration. The ebb of sky blue curve and the brown curve corresponding to time point 45 min and 1 h respectively, after A23187 administration have a gap; indicating stability of the process of cell breakdown may require further time

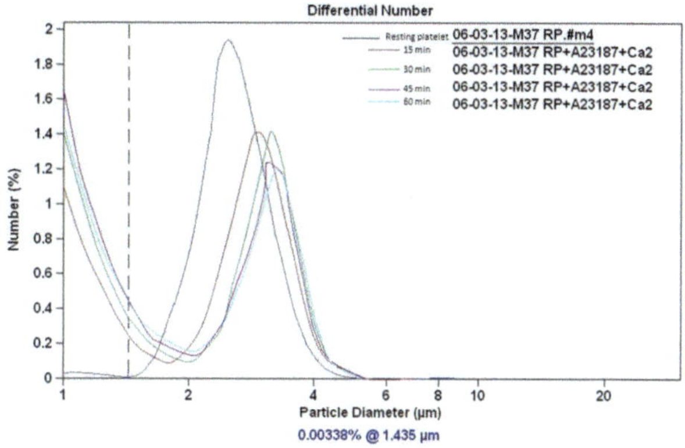

Fig. 13 Density of platelet size with A23187 (M 37, 06-03-13). We have seen the platelet size distribution may not attain stability within an hour after A23187 administration. We now consider longitudinal study on platelets under A23187 treatment for a different subject, a male aged 37 years, over a time period of 1 h recorded on 6 March 2013. This figure shows size distribution of platelets observed on 6 March 2013, when the chemical A23187 is added in platelet solution, and incubated for 5, 15, 30, 45, and 60 min, and the readings of the size distributions in number and percentage of cells versus size of cells are taken at the specified time. Deep blue curve corresponds to rest platelet distribution. The curves are drawn over increasing time gap from start at rest platelet (RP) position. The curves sequentially shift to the right, like in the case of earlier subject M 38, showing concentration of smaller sized platelets towards 1 μm diameter, indicating massive breakdown of the large platelets caused by A23187. Note the gradual shift of peak and ebb of the curves. The ebb coordinates (2.134, 0.116) of the pink curve and (2.158, 0.1715) of the sky blue curve corresponding to time point 45 min and 1 h, respectively, after A23187 administration show a gap between two points; indicating stability of the process of cell breakdown may require further time for this subject as well. See the abundance of small-sized particles near 1 μm, for curves other than the deep blue curve at the rest state, that has only 0.0038% of particles below the size 1.435 μm

Figure 15 shows six peak and ebb of mixture distribution for M 38 individual over a time range of 1 h. Spline regression indicates a possibility towards stability. In Fig. 16, the nine peak and ebb of mixture distributions are plotted for M 37 individual to examine the convergence to stability, where the last point(s) in a graphs refer to a time interval of 3 h. Both the curves after spline smoothing seem to stabilize at some level indicating the mixture distribution of blood platelets stabilizes around 3 h. The concentration of the chemical A23187 used is 1 micromole in the experiments. Distance between the peak and ebb in the sequence of time-dependent densities converges to a constant over time, indicating stability for large time gap, as shown in Figs. 17 and 18. Three-dimensional pictures of the variables are shown in Figs. 19, 20, 21, 22, 23, 24, and 25. The main characteristics of the bimodal density function arising out of this study are analyzed with the help of three-dimensional plots.

Details of the results are explained via the figures and the techniques of analysis are described in the legends.

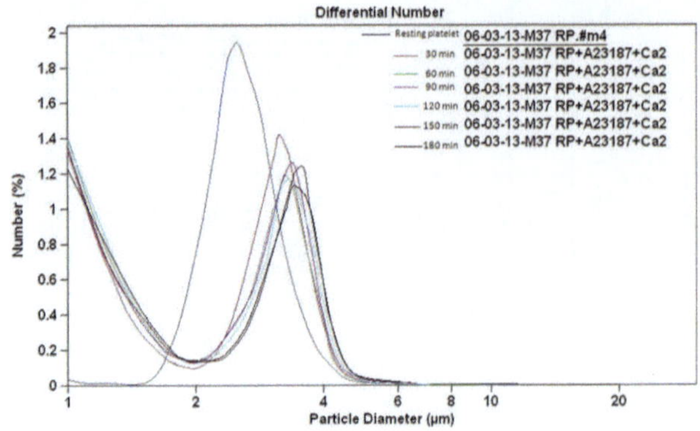

Fig. 14 Density of platelet size with A23187 (M 37, 06-03-13). We have checked again that the platelet size distribution may not attain stability in an hour after A23187 administration. We continue taking further observations over time to look for stability of the process in earlier experiment made for Fig. 13. The same experiment is continued for 3 h and records are taken intermittently. Coordinates of ebb in last two curves for 150 min and 180 min respectively are (2.132, 0.135) and (2.257, 0.173). We shall analyze the six peak and ebb of the curves shown in Fig. 12 for M 38 individual, and nine peak and ebb of the curves shown in Fig. 13 and this figure for M 37 individual to search for stability in the process

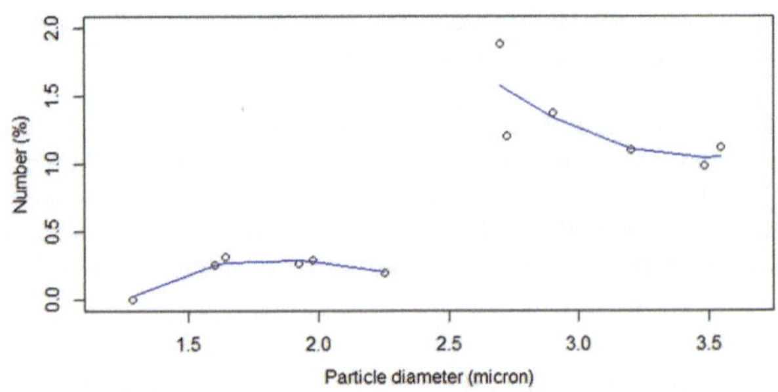

Fig. 15 Peaks and ebbs of platelet densities (M 38, ref. Fig. 12). In this figure, we plot the six peaks and ebbs of the curves shown in Fig. 12 in terms of (particle diameter, number %) coordinates. These appear in the top right corner and bottom left corner of this figure. Spline regression with d.f = 3.3 in each case shows that curves tend to be stable in y coordinate of number % towards the right end

4 A Growth Model

The Multisizer 4 machine can detect particles with diameter 1 μm and above. Starting with a single platelet at time $t = 0$, consider its break down at next time point; let there be n_{10} particles with diameter less than 1 μm with probability p_0, i.e., in term of birth and death process individual dies with probability p_0; n_{11} particles with diameter in $[1, 2)$ with probability p_1, n_{12} particles with diameter in $[2, 3)$ with probability p_3, etc. Thus, the expected number of particles at first stage is $\nu_1 = \sum_{j=1}^{\infty} p_j n_{1j}$. Consider a system where p_j are such that, from a single individual at stage 0, there

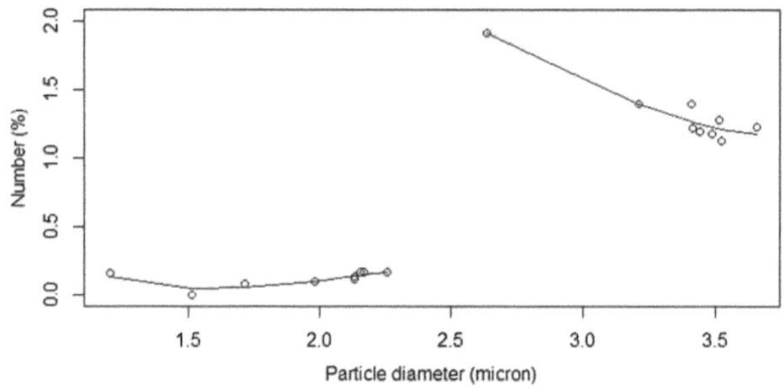

Fig. 16 Peaks and ebbs of platelet densities (M 37, ref. Figs. 13 and 14). There are nine points of peaks, and nine points of ebbs combining distinct time points in Figs. 13 and 14. Spline regression with d.f= 3.3 in each case shows that curves tend to be stable in y coordinate of number % towards the right end in this figure

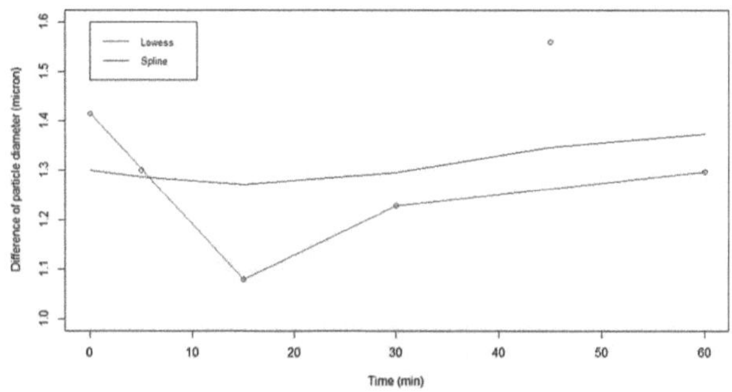

Fig. 17 Difference of particle diameter at peak & ebb of density versus time of recording (M 38, ref. Fig. 15). In order to look for stability in the process, for each curve in Fig. 12, for the individual M 38, at different time point of records, we compute the difference of x coordinates, i.e., difference in particle diameter at peak and ebb of the curves over time. When stability is achieved, differences should converge to a constant. The points are expected to lie in a line parallel to x axis as time increases. This figure shows six such points with lowess ($f = 2/3$) and spline regression (df $= 2.5$) versus time. Both lowess and spline regression indicate process approaches towards stability, as the slope of increase is low towards the right end

is at most one addition of individual to the process in the next stage on an average, and so on. In other words, proportionate growth is less than 2 at each stage of splitting in the population. Writing $y(t)$ as population size at time t, we then have $y(t+1)/y(t) = (1 + c_{t+1})$, where $|c| < 1$. Thus, $\log y(t+n) - \log y(t) = \sum_{i=1}^{n} \log(1 + c_{t+i}) \approx \sum_{i=1}^{n} c_{t+i} = \delta_{t,n} \to \delta$, as $n \to \infty$, if the series converges. Splitting rate at unit time is at most two, and that too slows down as the sum converges to a constant over time, so as to exclude explosive processes, these are ruled out in the present case.

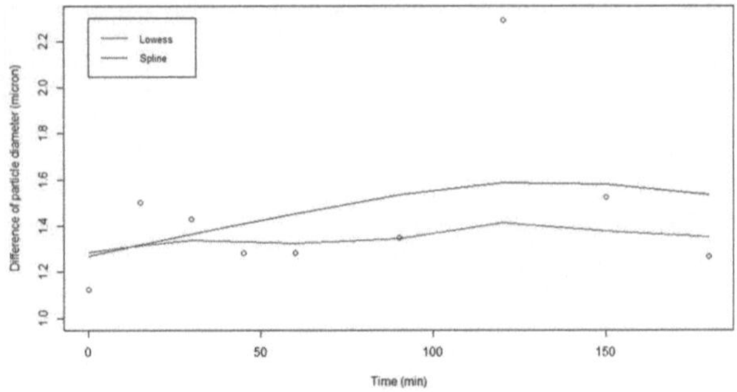

Fig. 18 Difference of particle diameter at peak and ebb of density versus time of recording (M 37, ref. Fig. 16). Consider the individual M 37. In order to look for stability in the process, for each curve in Figs. 13 and 14 at different time points of records, we compute the difference of x coordinates, i.e., difference in particle diameter at peak and ebb of the curves over time. When stability is achieved, the points are expected to lie in a line parallel to x axis as time increases. This figure shows nine such distinct points with lowess ($f = 2/3$) and spline regression (df = 2.5) versus time. Both lowess and spline regression at the right end indicates process approaches towards stability

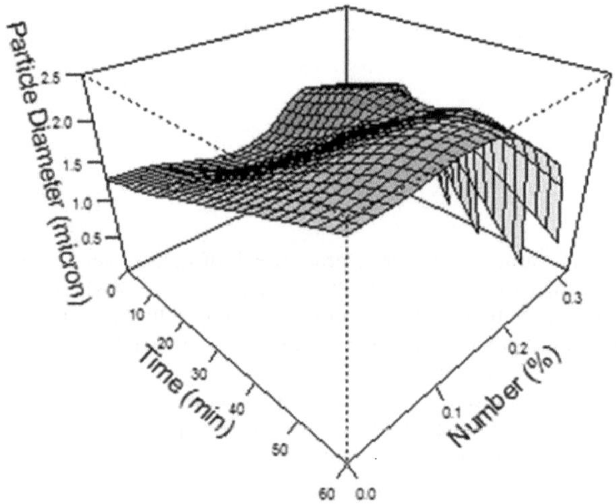

Fig. 19 Minimum particle diameter versus time and number (M 38, ref. Fig. 15). In order to check the stability of the process, we represent the relevant variables in a three-dimensional picture. Figure 15 relates to the subject M 38. The variables minimum density of particle diameter, i.e., the ebb of the density; number (%) were considered therein. We further consider the variable time of recording data with these two variables and draw a three-dimensional picture in R by the program "persp" in this figure. Stability is seen to achieve in the front-most corner of the picture with high value of time, low value of number (%), and particle diameter near 2.2 μm

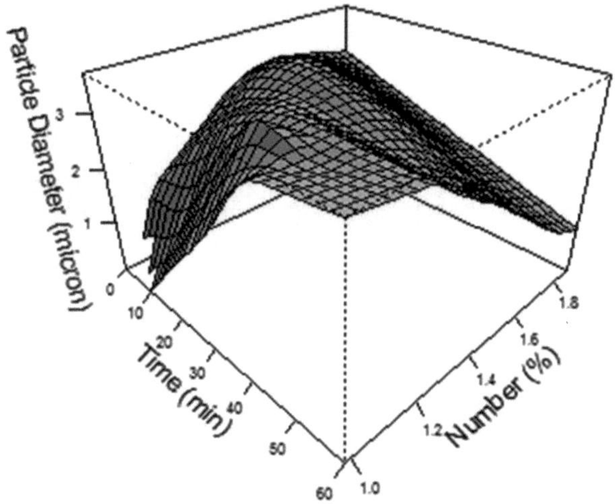

Fig. 20 Maximum particle diameter versus time and number (M 38, ref. Fig. 15). We plot maximum density of particle diameter, the peak of the density; to check the stability of the process, in a three-dimensional picture. Figure 15 relates to the subject M 38. The variables maximum density of particle diameter, i.e., the peak of the density; number (%) were considered therein. As in Fig. 19, we further consider the variable time of recording data with these two variables and draw a three-dimensional picture in this figure. Stability is seen to achieve in the front-most corner of the picture with high value of time, value of number 1.1 (%), and particle diameter near 3.7 μm

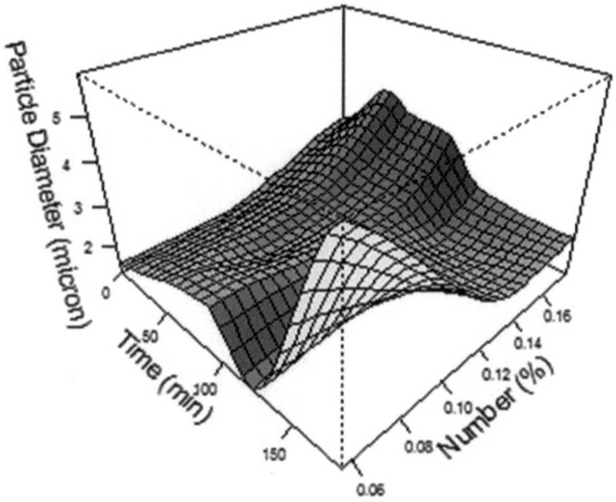

Fig. 21 Minimum particle diameter versus time and number (M 37, ref. Fig. 16). This figure related to M 37 is drawn in the same manner like Fig. 19. We check the stability of the process in a three-dimensional picture. Figure 16 relates to the subject M 37. The variables minimum (density of) particle diameter, i.e., the ebb of the density; number (%) are considered therein. We consider the variable time of recording data with these two variables also and draw a three-dimensional picture in R by the program "persp". Stability is seen to achieve in the distant corner in right-hand side of the picture with high value of time, value of number (%) around 0.17, and particle diameter near 2.2 μm

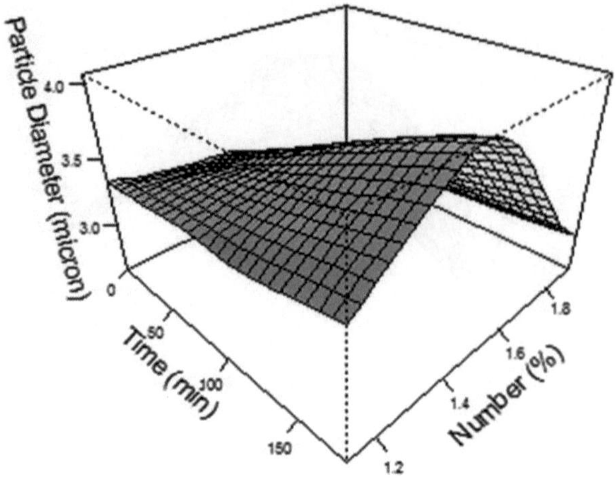

Fig. 22 Maximum particle diameter versus time and number (M 37, ref. Fig. 16). This figure related to M 37 is drawn in a similar manner like Fig. 20 to check the stability of the process in a three-dimensional picture. Figure 16 relates to the subject M 37. The variables maximum (density of) particle diameter, i.e., the peak of the density; number (%) were considered therein. We consider time of recording data with these two variables. Stability is seen to achieve in the middle elevated portion in right-hand side of the picture with high value of time at 3 h, value of number (%) around 1.6, and particle diameter near 4 μm

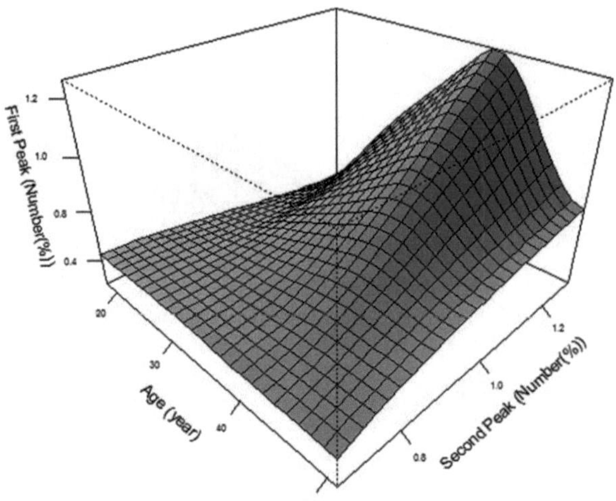

Fig. 23 Small size particle accumulation after 15 min under A23187: first peak versus age and second peak. Consider Figs. 3, 5, 6, 7, 8, 9, 10 and 11, showing particle concentration after 15 min under A23187 administration to individuals of different ages. In Fig. 23, we study the variation of small size particle concentration in the first peak, i.e., we consider height of the density at first peak with respect to age of individuals and height of the density at the second peak. It appears that within the age range of 37–38.5 years; the two peaks attain the maxima simultaneously. The surface drops down towards ground level zero smoothly to other sides. Increase in the height of first peak is seen with increase in second peak for fixed age beyond 40 years

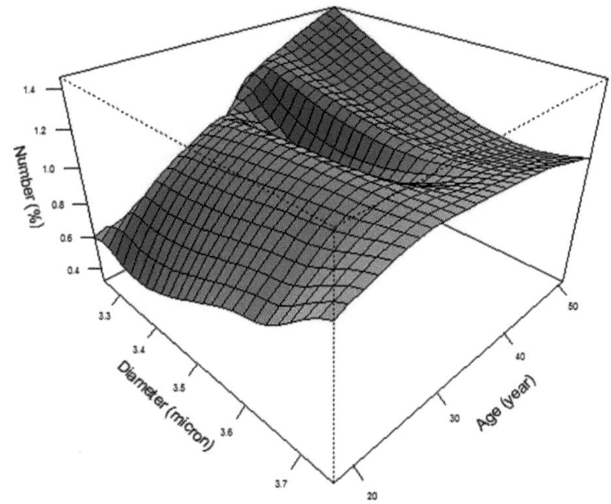

Fig. 24 Small size particle accumulation after 15 min under A23187: first peak versus age and diameter at second peak. For Figs. 3, Figs. 5, 6, 7, 8, 9, 10, and 11, the diameter of the particles in first peak is 1 μ_m, but the diameter at second peak varies over age. In this figure, we consider height of the density at first peak with respect to diameter of the particles at second peak and age of individuals. A drop in the height of surface is seen for age of individuals at about 45 years, density height of the small size particles at first peak is low at that age, especially when particle diameter is below 3.6 μm, at the second peak. For age below 30 years, the surface is slanting downward, indicating fall in the height of density at 1 μm, in general for all values of particle diameter at second peak

In such a situation, the size of the population; here the number of particles starting from time t, after a time gap of n units is $e^{\delta_{t,n}} y(t) \approx e^{\delta} y(t)$, indicating stability of the process with proportionate growth.

The model suggests a cap on the rate of breakdown in splitting process of platelets, so as to reach stability in the long run. This is consistent with the observed phenomena and provides an interpretation of the proportional growth e^{δ} under stability in time, in terms of successive breakdown process over time.

5 Discussions

We study the distributional properties of platelet size breakdown affected by the chemical A23187 with concentration 1 micromole; A23187 is antibiotic against Gram-positive bacteria and fungi. This acts as a divalent cation ionophore, see, e.g., Bernard et al. (1979), allowing the ions to cross cell membranes, which are usually impermeable to the ions. Blood platelet size distribution is affected by the chemical A23187, as this induces breakdown of platelets to smaller size. The rest platelet distribution is unimodal. Under A23187 administration, the distribution becomes

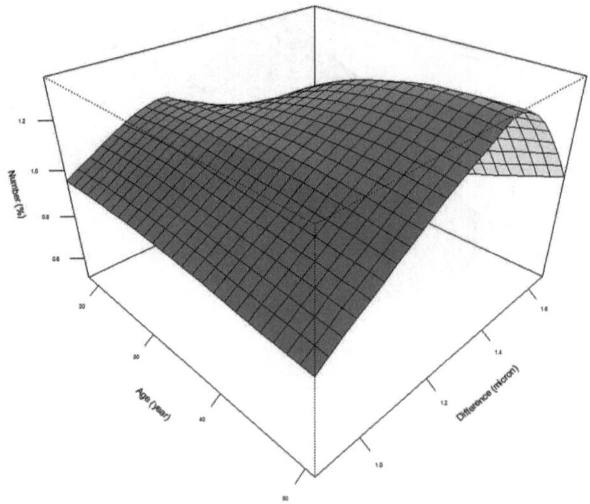

Fig. 25 Small size particle accumulation after 15 min under A23187: first peak versus age and difference of diameters at second peak and ebb. We also consider the behavior of diameter at ebb for Figs. 3, 5, 6, 7, 8, 9, 10, and 11 over age. In Fig. 25, small size particle accumulation after 15 min of A23187 administration to the platelet solution is examined at the first peak with respect to age and difference of diameters at second peak and the ebb, which is located in between first and second peak. A general trend of increasing height of particle density at first peak is observed over all age with increase in the difference of diameters at second peak and ebb up to 1.3 μm. The same phenomenon of increasing the height of first peak is observed, for age higher than 30 when difference of diameters is less than 1.4 μm; after which downward slope of surface is seen towards the right-hand side of the Fig. 25. Topmost value of the surface is attained when difference of diameters is 1.4 μm; and age is greater than 40 years

bimodal. The splitting process is not explosive, but seems to be proportionate, and can be explained by a theoretical growth model. Distributions vary over different individuals of different ages, as observed in three-dimensional Figs. 23, 24 and 25. Stability of the size breakdown is observed at a gap of about 3 h from A23187 administration. The study is of help in explaining drug carrier role of the chemical to cells.

References

Bernard, A., Fukuda, D. S., Dorman, D. E., Occolowitz, J. L., Debono, M., & Farhner, L. (1979). Microbial transformation of A23187, a divalent cation ionophore antibiotic. *Antimicrob Agents Chemotherapy*, *16*(6), 808–812.

Cleveland, W. S. (1981). LOWESS: A program for smoothing scatterplots by robust locally weighted regression. *The American Statistician*, *35*(1), 54.

Feller, W. (1939). Die Grundlagen der Volterraschen Theorie des Kampfes ums Dasein in wahrscheinlichkeitstheoretischer Behandlung. *Acta Biotheoretica*, *5*, 11–40.

Gerrard, J. M., White, J. G., & Rao, G. H. R. (1974). Effects of the Ionophore A23187 on Blood Platelets II. Influence on Ultrastructure. *The American Journal of Pathology*, *77*(2), 151–166.

Kendall, D. G. (1948). On the generalized birth and death process. *The Annals of Mathematical Statistics*, 1–15.

Khinchin, A. Ya. (1938). On unimodal distributions. Trams. Research Institute of Mathematics and Mechanics (in Russian). University of Tomsk. *2*(2), 1–7.

Mokaddis, G. S. (1985). A generalized age-dependent branching process and its limit distribution. *Indian Journal of Pure and Applied Mathematics*, *16*(12), 1399–1406.

Reinsch, C. H. (1967). Smoothing by spline functions. *Numerische Mathematik, 10*, 177–183. https://doi.org/10.1007/BF02162161.

Stratton, H. H., & Tucker, H. G. (1964). Limit distributions of a branching stochastic process. *The Annals of Mathematical Statistics*, *35*, 557–565.

Tardi, P., Choice, E., Masin, D., Redelmeier, T., Bally, M., & Madden, T. D. (2000). Liposomal encapsulation of topotecan enhances anticancer efficacy in Murine and Human Xenograft models. *Cancer Research*, *60*(13), 3389–3393.

White, J. G., Rao, G. H. R., & Gerrard, J. M. (1974). Effects of the ionophore A23187 on blood platelets. I. Influence on aggregation and secretion. *The American Journal of Pathology*, *77*, 135–150.

Correction to: A Statistical Study of Platelet Size Decomposition and Related Growth Model

Ratan Dasgupta, Koushik Biswas and Debabrata Dash

Correction to:
Chapter "A Statistical Study of Platelet Size Decomposition and Related Growth Model" in: R. Dasgupta (ed.), *Advances in Growth Curve and Structural Equation Modeling*, https://doi.org/10.1007/978-981-13-0980-9_11

The original version of the chapter was inadvertently published without authors' names and their affiliations, Dr. Koushik Biswas, Regional Institute of Ophthalmology, Medical College Hospital, Kolkata-700073, India. Email: koushik2907@gmail.com and Dr. Debabrata Dash, Department of Biochemistry, Institute of Medical Sciences, Banaras Hindu University, Varanasi 221005, India. Email: ddash.biochem@gmail.com in the opening page of chapter "A Statistical Study of Platelet Size Decomposition and Related Growth Model", which have now been included. The correction chapter and the book have been updated with the changes.

The updated version of this chapter can be found at
https://doi.org/10.1007/978-981-13-0980-9_11

R. Dasgupta (ed.), *Advances in Growth Curve and Structural Equation Modeling*,
https://doi.org/10.1007/978-981-13-0980-9_12

Some Snapshots from the Workshop on Growth Curve Model, held during 23–24 February 2017 at the Indian Statistical Institute, Giridih

Picture 1 Garlanding the statue of P. C. Mahalanobis, Founder of the Indian Statistical Institute; before commencement of the workshop 2017

R. Dasgupta (ed.), *Advances in Growth Curve and Structural Equation Modeling*,
https://doi.org/10.1007/978-981-13-0980-9

Picture 2 Growth curve model 2017 workshop participants and workers of the Indian Statistical Institute, Giridih

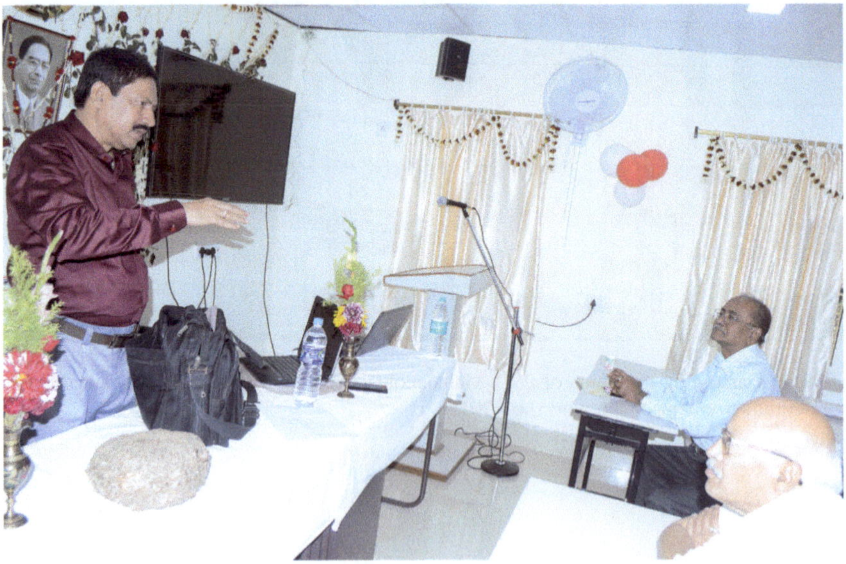

Picture 3 Workshop lectures

Picture 4 Tribal students of Damcherra high school, Tripura

Picture 5 Tribal students are going to a primary school in the interior of Tripura, near a tea garden

Picture 6 Tribal girls in a school of Tripura

Picture 7 Weekly market in a tribal area of Tripura

Picture 8 Tribal boys in a school of Tripura

Picture 9 Cultivated potato tubers attached to stems

Picture 10 Elephant foot yam cultivation in shade, to protect the plants from harsh sunlight of Giridih summer

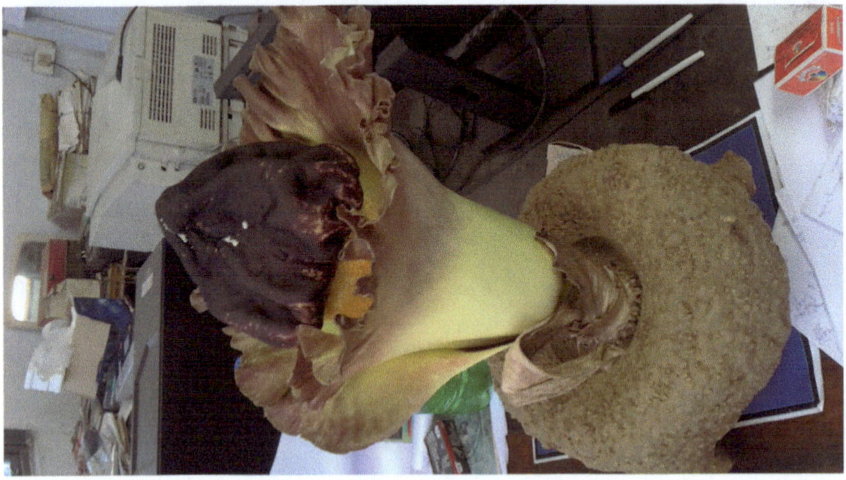

Picture 11 Elephant foot yam grown in Giridih farm, photographed later in Kolkata office, in a blooming state

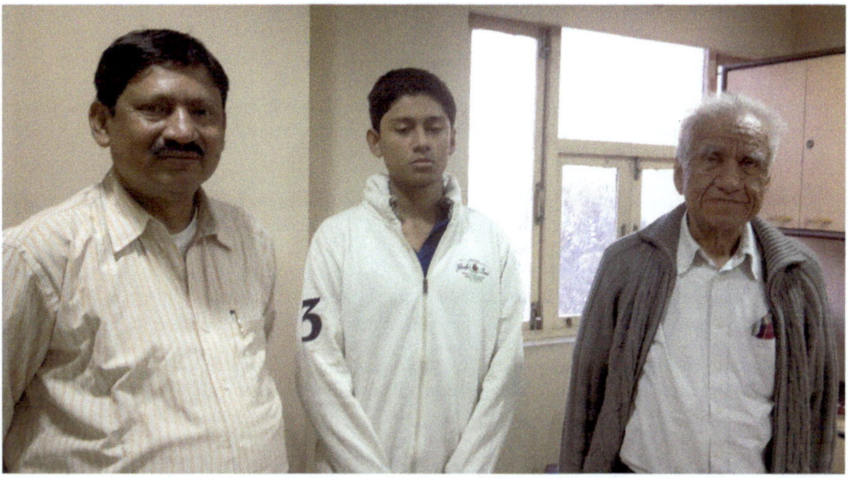

Picture 13 Professor Jayanta K. Ghosh in his office with editor on January 12, 2015. From right to left: J. K. Ghosh, Debkumar, Editor

Director, ISI

May I make a request on behalf of Prof. Ratan Dasgupta.

As you know Ratan is both a Probabilist and a Statistician who can effortlessly move, from one area of research into another and combines the goals of welfare for marginal people in India with hard theoretical work. Such people are very rare.

His current and new proposed work addresses welfare of tribals, women, and deprived children with inadequate access to good learning aids for math, and adult in neglected areas, who are often cheated by sophisticated money lenders.

His mission is the same as that of most respected people in the world who help children, tribals and other deprived people.

Today he lectured about some of his goals and methods to me and Professor Partha Majumdar, the renowned human geneticist. Both of us quizzed him a lot and feel satisfied by both his goals and methodology of research.

Could ISI provide about Rs 5 lac? If that seems too much I am willing to bear a part the expenditure as part of my global interest and urge to help vulnerable people namely women and children.

Ratan has three areas of interest

1) Growth curve model experiments in ISI, Giridih
2) Experimental work at Sunderban saline soil on yam and coconut trees
3) Work among north East tribes on development and aptitude.

Could some fund be made available from ISI for continuation of work for the next financial year starting on 1st April, 2016.

I would greatly appreciate your support and help to this noble cause.

We are passing through a rather bleak period of neglected children and women all over the world.

Whatever little we can do will matter a lot for at least our own country. I would be willing to provide some support even though I am now a very old man – eighty years old.

Best regards,

J K Ghosh

(J K Ghosh)
Emeritus Prof ISI, Purdue University

24th March, 2016

Picture 14 Professor J.K.Ghosh always encouraged theory-cum-applied research

MIX
Papier aus verantwortungsvollen Quellen
Paper from responsible sources
FSC® C105338

If you have any concerns about our products,
you can contact us on
ProductSafety@springernature.com

In case Publisher is established outside the EU,
the EU authorized representative is:
Springer Nature Customer Service Center GmbH
Europaplatz 3, 69115 Heidelberg, Germany

Printed by Libri Plureos GmbH
in Hamburg, Germany